A Primer on Environmental
Policy Design

FUNDAMENTALS OF PURE AND APPLIED ECONOMICS

EDITORS IN CHIEF

J. LESOURNE, Conservatoire National des Arts et Métiers, Paris, France
H. SONNENSCHEIN, School of Arts and Sciences, Philadelphia, Pennsylvania, USA

ADVISORY BOARD

K.ARROW, Stanford, California, USA
W. BAUMOL, Princeton, New Jersey, USA
W. A. LEWIS, Princeton, New Jersey, USA
S. TSURU, Tokyo, Japan

The Fundamentals Series' sections and editors, and published titles may be found at the back of this volume.

Titles in the Government Ownership and Regulation of Economic Activity section

A Primer on Environmental Policy Design

by Robert W. Hahn
Council of Economic Advisers
Washington, DC.

A volume in the Government Ownership and
Regulation of Economic Activity section

edited by

E. Bailey
Carnegie–Mellon University
Pittsburgh, Pennsylvania

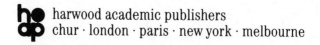
harwood academic publishers
chur · london · paris · new york · melbourne

© 1989 by Harwood Academic Publishers GmbH
Poststrasse 22, 7000 Chur, Switzerland
All rights reserved

Harwood Academic Publishers

Post Office Box 197	58, rue Lhomond
London WC2E 9PX	75005 Paris
England	France

Post Office Box 786	Private Bag 8
Cooper Station	Camberwell, Victoria 3124
New York, NY 10276	Australia
United States of America	

Library of Congress Cataloging-in-Publication Data

Hahn, Robert William.
  A primer on environmental policy design/Robert W. Hahn.
  p. cm.—(Fundamentals of pure and applied economics; v. 34.
Government ownership and regulation of economic activity section)
  Bibliography: p.
  Includes index.
  ISBN 3-7186-4897-0
  1. Environmental policy. I. Title. II. Series: Fundamentals of
pure and applied economics; v. 34. III. Series: Fundamentals of
pure and applied economics. Government ownership and regulation
of economic activity section.
HC79.E5H316   1989
363.7—dc 19   88-38992
                                                            CIP

CONTENTS

Introduction to the Series

Drawing on a personal network, an economist can still relatively easily stay well-informed in the narrow field in which he works, but to keep up with the development of economics as a whole is a much more formidable challenge. Economists are confronted with difficulties associated with the rapid development of their discipline. There is a risk of "balkanization" in economics, which may not be favorable to its development.

Fundamentals of Pure and Applied Economics has been created to meet this problem. The discipline of economics has been subdivided into sections (listed inside). These sections include short books, each surveying the state of the art in a given area.

Each book starts with the basic elements and goes as far as the most advanced results. Each should be useful to professors needing material for lectures, to graduate students looking for a global view of a particular subject, to professional economists wishing to keep up with the development of their science, and to researchers seeking convenient information on questions that incidentally appear in their work.

Each book is thus a presentation of the state of the art in a particular field rather than a step-by-step analysis of the development of the literature. Each is a high-level presentation but accessible to anyone with a solid background in economics, whether engaged in business, government, international organizations, teaching, or research in related fields.

Three aspects of *Fundamentals of Pure and Applied Economics* should be emphasized:

—First, the project covers the whole field of economics, not only theoretical or mathematical economics.

—Second, the project is open-ended and the number of books is not predetermined. If new and interesting areas appear, they will generate additional books.

—Last, all the books making up each section will later be grouped to constitute one or several volumes of an Encyclopedia of Economics.

The editors of the sections are outstanding economists who have selected as authors for the series some of the finest specialists in the world.

J. Lesourne   *H. Sonnenschein*

Acknowledgements

This book has been inside of me for a long time. Several people were instrumental in helping to allow it emerge in a form which is accessible to others (i.e., the written word). I would like to thank Betsy Bailey for reviewing the entire manuscript and providing helpful comments. In addition, I would like to thank Roger Noll for providing helpful comments on various parts of the manuscript.

Some of the material in this book has already been published in academic journals. Section 3 builds on an article in the *Journal of Economic Perspectives*. A preliminary version of Section 5 appeared in the *Natural Resources Journal*. I am grateful to these journals for granting permission to use this material in compiling this book.

Funding for this book was provided in large part by the Decision Risk and Management Science Program at the National Science Foundation. This support not only made my life easier, but also helped increase my productivity. In deference to my financial supporters, I hasten to add that the final work is ultimately my responsibility.

Having ascribed responsibility for legal purposes, let me now attempt to apportion the remainder of the credit. The manuscript would have been much more difficult to write had it not been for some of my earlier joint work with Gordon Hester. Undoubtedly, many of the ideas in this manuscript were jointly stumbled upon by Gordon and myself.

For help in actually getting and keeping the project going, I would like to thank Pam Madden and Judy Schachter, both of whom aided in unquantifiable, but nonetheless important, ways. Donna Dowd and JoAnn Kline were instrumental in helping to produce the final form of the manuscript.

Finally, I would like to thank Greg McRae for never failing to remind me of the adage of one of our most venerable Caltech professors, Joel Franklin. It goes, "A page a day is a book a year." And so emerged what you have before you.

Preface

This is a book about characterizing and solving environmental problems. It examines the design of environmental policy in the "real world." The purpose is to help provide a framework to better understand and evaluate environmental policy. The book is written for anyone interested in understanding what economists and political scientists have to add to the debate over environmental policy. It is written in the form of an unfinished symphony. My purpose in writing the book is to provide an assessment of the current state of the art and to offer directions for future research.

The book is designed to be accessible to a wide audience. While some exposure to microeconomics and basic political science would be helpful, this is not essential. The only material of a highly technical nature is contained in Section 4. However, the basic ideas should be accessible even if the formal mathematics is not. The remaining chapters occasionally make use of simple graphical concepts, but otherwise are not technically demanding.

A Primer on Environmental Policy Design

ROBERT W. HAHN

Carnegie Mellon University, Pittsburgh, Pennsylvania and Council of Economic Advisers, Executive Office of the President, Washington, D.C.

1. INTRODUCTION

As a child, I was briefly influenced by the recycling "movement" that had achieved some degree of popularity in middle class suburbs throughout America during the late sixties. Many people were concerned that too little attention was being given to the environment around us. To join the vanguard, I asked my parents to recycle bottles and aluminum cans at the local recycling center. Initially, things worked quite well from my point of view. My sister and I would put the bottles and cans into plastic bags and accompany my father on a trip to the center, which was located by the beach. As time went by, however, I must confess that my enthusiasm waned, and I lost interest in recycling. My initial response was to urge my father to continue while I pursued other more worldly pleasures such as basketball. For a while, this strategy worked, but eventually my father decided to pursue his worldly vice, which was (and still is) golf. At some point, I noticed that Dad was not taking recycling seriously anymore, and mentioned this fact to him in what must have been an obnoxious manner. His response, which I will never forget, was "Ecology starts in the home—clean up your room." And thus began my career as a student of environmental science and policy.

The lessons from my experience as a child have had a great impact on the way I think about environmental policy. Most importantly, I learned the importance of motivation and staying power. As the initial excitement over the recycling movement was replaced by the drudgery of transporting cans, it became clear to me that my contribution to solving the problem was really only a drop in the bucket. That wouldn't have mattered to me much if there were something there to keep my continued interest, but there wasn't. That was a very important lesson. I learned that the system of economic and social rewards associated with a policy can have a major impact on how a person behaves in different situations. Finally, there was the unforgettable response of my father, who was obviously fed up with the whole ordeal. That taught me that people often have divergent interests, and will go to great lengths to move policy in a direction that will be useful for them.

The trick would appear to be to design ways to achieve socially desirable goals while allowing people to pursue their individual self-interest. This is no small feat. My objective in this book is to examine approaches to this problem in designing environmental policy. The framework for the book builds on the disciplines of political science and economics. Both disciplines are unified in their focus on the self-interest of the individual. Political science focuses on the activities of individuals in the political sphere. Economics focuses on the interests of individuals in managing their own resources. In reality, the two problems are inseparable. Politics can have a significant impact on the opportunities available to an individual to pursue economic gain. Thus, it makes sense to integrate the two notions of self-interest under a single roof. This, in essence, is what political economy is all about.

This book is an attempt to integrate our economic understanding of environmental policies with our understanding of political institutions. To my knowledge this is the first attempt to do so in a volume which reflects the current state-of-the-art in both of these disciplines. Section 2 provides an overview of environmental policies, and also examines the basic frameworks used by economists and political scientists to address environmental problems. Actual experience in the use of new approaches to environmental management is chronicled in Section 3. The section focuses on the actual application of emission fees and marketable permits, two instruments that have received widespread support from economists. Section 4 offers some explanations for a variety of patterns that are observed in environmental regulation. The problem of designing systems that improve environmental quality and increase productivity is addressed in Section 5 in the context of a particular pollution problem. The case study, which identifies new approaches for helping to meet the current U.S. ozone standard, is helpful in identifying how general insights about the performance of regulatory systems can be applied to specific problems. Finally, Section 6 highlights the principal conclusions and suggests areas for future research.

2. FRAMEWORKS FOR ANALYZING ENVIRONMENTAL POLICY

2.1. Introduction

Public concern over environmental issues has risen dramatically over the last two decades. There has been increasing pressure on the governments of industrialized countries to develop a reasoned response to a variety of problems, such as acid rain, the clean-up of toxic waste dumps and urban air pollution. The people in charge of developing and administering these control programs face a gargantuan task. To help make this task more manageable, policy analysts have developed a series of models for understanding the implications of pursuing different environmental control strategies. The purpose of this section is to provide an overview of selected environmental approaches to problem solving and to suggest frameworks for evaluating and understanding these approaches.

The primary aim of this book will be to serve as an aid in the design of less costly, and more effective, policies for addressing environmental problems. If the need for identifying such policies is not self-evident, a review of environmental expenditures over time helps to provide a rationale. Figure 1 lists pollution control expenditures in the United States from 1972 to 1986 (Farber and Rutledge, 1986, 1987). Nominal expenditures have risen dramatically over this period from $18 billion to $78 billion. Real expenditures have also increased by over 60 percent during this period. Currently, about 2 percent of the entire gross national product is used to address environmental problems. Expenditures will probably continue to rise for the foreseeable future. Heightened concern over a variety of problems ranging from the depletion of stratospheric ozone to the generation and disposal of hazardous wastes will give rise to regulations with very sobering price tags. For example, a reauthorization bill proposed for cleaning up hazardous

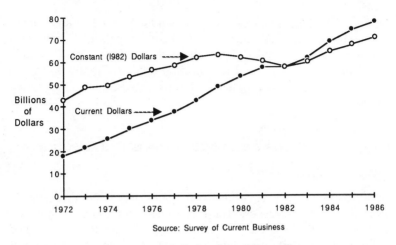

Source: Survey of Current Business

FIGURE 1   U.S. Pollution Control Expenditures

waste in the U.S. would cost taxpayers $9 billion (*Wall Street Journal*, 1986).

Simply stated, the central issue addressed in this book is whether it is possible to get more for less. While apparently a simple question, it defies a simple answer. In theory, it is often easy to design systems that will give you "more bang for your buck." The real world is quite another story, however. In most cases, executing policies in the real world is much more complex than theory admits. Moreover, there are frequently very strong political forces that constrain what can and cannot be done.

One measure of the strength of political forces is the fraction of expenditures that are earmarked for pet projects of Congressmen (i.e., "pork barrel" projects). The precise definition of such projects is always problematic. Two programs that are thought to have a large pork barrel component are the Construction Grants Program and Superfund (Arnold, 1979; Smith, 1986). The Construction Grants Program is used to help finance sewage treatment plants. Superfund is used to help clean up old hazardous waste sites, many of which have been abandoned. Figure 2 shows the percentage of EPA funds allocated to both of these programs. In 11 of

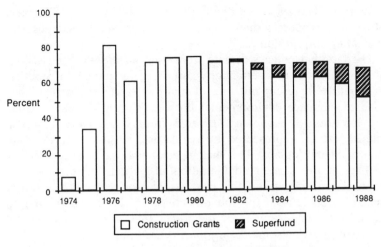

Source: Office of Management and Budget

FIGURE 2   Construction Grants and Superfund as a Percentage of EPA Budget

the last 13 years, these two programs accounted for over 70 percent of the total EPA budget, with the lion's share going to Construction Grants. In the future, this balance can be expected to change as Superfund gains increasing support. The figure provides striking testimony to the fact that much of environmental policy has its roots in electoral politics.

Because politics has a strong impact on the shape of environmental policy, it should come as no surprise that programs are rarely designed with economic efficiency as a primary goal. For many years, economists have argued that it is indeed possible to get more for less by placing greater reliance on the use of economc incentive mechanisms in protecting the environment. While many of these arguments were compelling in theory, they lacked the force that practical examples can offer. As we shall see in Section 3, a review of practical experience reveals that economic applications deviate dramatically from standard textbook discussions of these approaches. Nonetheless, in certain situations, approaches based on economic theories have had dramatic effects on efficiency. Moreover, economic theory can serve as a useful guidepost for evaluating current policies.

The fact that policies deviate from some of the idealized visions of economists is not particularly surprising to students of politics. Economists represent but one of many interest groups that have tried to influence environmental problem solving. In many cases, the interests of economists diverge from other more powerful interest groups represented by regulators, business interests and environmental lobbies. The challenge for political science is to try to explain the current configuration of policies on the basis of political institutions and the interests of key actors in the policy making process. One of the objectives of this book will be to extend our understanding of the forces that shape environmental policy by applying some basic concepts used in political science.

A useful starting point for developing more effective policy is to characterize normative frameworks that allow us to compare the performance of different approaches. This is done in part 2.2 of this Section. Part 2.3 then provides an overview of various policy instruments and examines their theoretical performance charac- teristics. The political motivations of regulators and legislators for choosing various policies are explored in part 2.4. Finally, part 2.5 summarizes the key ideas presented in the chapter.

2.2. The Normative Framework

The focus of this book will be on designing policies that have the potential to meet environmental objectives in a more timely and efficient manner. Frameworks can be useful as a tool for helping to guide design. Two basic frameworks will be employed here. The first, borrowed from welfare economics, is normative. It provides measures for evaluating the efficiency and effectiveness of different policies. The second, borrowed from political science, is positive. It identifies key factors that affect how policies are implemented and which policies are likely to be feasible.

The basic normative framework used in applied welfare econom- ics is designed to help examine the relative efficiency of different policies. The hallmark of the normative approach is to specify a desired objective. A typical objective is to identify policies that

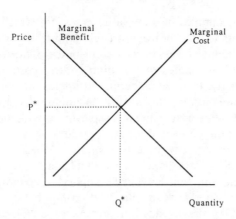

FIGURE 3 The Standard Paradigm in Microeconomics

maximize net benefits. An efficient policy is defined as one which maximizes net benefits. This idea is illustrated in Figure 3, which shows the marginal costs and benefits associated with various levels of pollution abatement. When marginal benefit and cost curves are well-behaved, as shown in the figure, the problem is to identify the point (P^*, Q^*) and then identify control strategies accordingly. While this paradigm provides a useful starting point, it is usually difficult to implement in a straightforward manner. There are great uncertainties in identifying marginal benefits and costs.

To understand the importance of uncertainty in this formulation of the problem, it is instructive to examine the origins of the marginal benefit and the marginal cost curves shown in Figure 3.[1] For environmental policy problems, analysis of the net benefits of control requires estimating both the benefits and costs of control as a function of the quantity of emissions of various pollutants. The first step in the analysis is to identify which pollutants to control. The answer may be clear, as in situations where a chemically inert pollutant is emitted from a single "pipe" into an isolated environmental reservoir (some lakes, for example). In other cases, numerous sources emit complex mixtures of reactive species into nonisolated airsheds and watersheds. The acid deposition and urban ozone

---

[1] The part of this section dealing with uncertainty is drawn from Hahn, McRae, and Milford (1985).

problems are examples of these latter cases, for which the question of which agents to control has not been definitively answered.

Once the chemical agents that contribute to the problem have been identified, estimates of pollution control costs can either be obtained directly from engineering estimates or indirectly from estimated production functions. Both techniques have strengths and weaknesses. Engineering estimates typically provide detailed estimates of pollution control options, but little information on the uncertainty associated with such estimates. Moreover, they rarely consider detailed process changes that could arise. Production function estimation, as it is typically applied in econometrics, tends to focus more on the aggregate relationship between inputs and outputs, treating the production process as a "black box." This estimation procedure provides estimates of the error associated with various parameters. However, such estimates may be of dubious value if the model does not adequately capture the structural relationship between inputs and outputs.

Estimation of the marginal benefits of different control strategies usually involves two steps: the first is the determination of the effect of the controls on environmental quality, and the second is the evaluation of the benefits that accrue from incremental changes in environmental quality. There are usually large uncertainties associated with both steps. For example, consider the problem of controlling acid deposition. The control measure usually suggested for reducing acid deposition is reduction of emissions of sulfur oxides from power plants and other industrial sources. The chemical and physical processes that occur between emission and deposition of sulfur species makes it difficult to accurately predict how much a given level of emissions control will reduce deposition. The pollutants may undergo both gas-phase and aqueous-phase chemical reactions, and be transported over distances up to hundreds of kilometers. Some of the chemical steps, especially those occurring in clouds, are not well understood (National Academy of Sciences, 1983). The meteorological conditions that affect both chemical transformations and physical transport of the pollutants vary from day-to-day and year-to-year, contributing to the uncertainty in estimates of the relationship between deposition and emissions control. Finally, just measuring the amount of sulfur that has deposited on various surfaces has proved to be an extremely difficult task.

Placing a dollar value on the damages associated with acid deposition is even more problematic. Benefits estimation is an inexact science at best. A variety of techniques are used to estimate the dollar value of changes in control levels. For example, in the case of materials damage to buildings, some estimates of the increased costs associated with maintaining buildings or using different construction materials may be used. To assess the effect that acid rain has on lakes and recreational fishing, survey data are used to estimate the opportunity cost of fishing in other lakes. In instances where acid deposition is thought to affect human health, regression models may be used to infer the increase or decrease in morbidity and mortality. These numbers then need to be converted into dollar values. It should be quite apparent from this brief examination of the acid deposition problem that the development of benefits techniques is still in its infancy. Virtually all existing benefits techniques are characterized by large uncertainties, sometimes involving several orders of magnitude (Freeman, 1979).

The question naturally arises as to how to deal with these uncertainties. Uncertainty in the nature of the problem needs to be dealt with by selecting policy instruments that are *flexible*. By defining policies in a way that allows them to accommodate changes in our understanding of the underlying science, resources can be saved (Hahn and Noll, 1982a, 1982b). Another way of saving resources is to design systems that provide firms greater flexibility in their choice of abatement options for meeting prescribed objectives.

Because of difficulties in estimating benefits, economists often compare policies in terms of their *cost-effectiveness*. A cost-effective policy is one which meets a prescribed objective at the lowest possible cost. In the case of environmental problems, this objective is sometimes stated in terms of an aggregate emissions target, such as the number of tons of sulfur oxides that can be emitted in the eastern United States in a given year.

2.3. The Selection of Instruments

In thinking about the design and implementation of policies, it is generally assumed that policy makers can choose from a variety of mechanisms for achieving specified objectives. These mechanisms are often referred to as "instruments." While the precise definition of an instrument is somewhat arbitrary, it suffices to think of an instrument as a particular method for achieving a result. Thus, for example, a town may choose between implementing a general sales tax or a property tax as a way of raising needed revenue. In this case, the two taxes can be thought of as different instruments.

2.3.1. The Regulator's Toolchest

This book will examine the selection of instruments by politicians and bureaucrats charged with addressing environmental problems. There are many different ways to induce firms to meet environmental objectives. For purposes of this study, they can be divided into the following four categories: standards, subsidies, taxes and marketable permits.[2]

Standards come in two general varieties. Ambient standards specify the overall level of environmental quality in a region or a waterway. Emissions standards specify emission limits on individual sources that emit pollutants. Emissions standards and ambient standards are usually related in the sense that regulatory agencies try to specify emissions standards that are consistent with ambient standards. They are not always successful, however. The relationship between emissions and, say, air quality is sometimes very difficult to predict with a high degree of certainty.

Here, we shall be primarily concerned with emissions standards, since it is these standards that directly affect the actions of firms. Emissions standards represent the dominant instrument in environmental regulation throughout the world. There are two very common types of emission standards: technology-based standards and performance standards. As the name implies technology-based standards identify a particular technology that must be used to comply with the regulation. For example, utilities may be required to use a scrubber to control sulfur oxide emissions. This type of standard is used frequently in both air and water pollution regulation in the United States. Performance standards are more flexible than technology-based standards. A performance standard typically defines a performance measure and allows firms to select the best way to meet this standard. Thus, for example, firms were asked to develop a technology for automobiles that resulted in emissions no greater than 0.41 grams of hydrocarbons, 3.4 grams of

---

[2] See Bohm and Russell (1985) for a more extended discussion.

carbon monoxide, and 1.0 grams of nitrogen oxides per mile. Almost all standards share the characteristic that the firm complying with the standard bears the initial cost; however, some or all of these costs are eventually passed on to consumers.

Subsidies are also widely used as a vehicle for promoting environmental quality in developed countries. Examples include subsidies for exploring particular types of environmentally benign technology, such as solar energy and wind, as well as subsidies for building sewage treatment plants. Subsidies take many forms. Some come in the form of tax credits, others in the form of below-market financing for pollution control equipment, and still others relate directly to the purchase of specific types of abatement equipment.

Taxes receive less frequent use than either subsidies or standards. In the environmental arena, the taxes that are most frequently used are *emissions fees*.[3] These fees are typically assessed based on the actual or expected emissions from different kinds of polluting activities. Recently, fees have enjoyed increasing popularity in Europe as a means for raising revenues which are then used to promote environmental quality. For many years, taxes were a favorite tool promoted by the economics community because they have some desirable efficiency properties. The idea underlying emissions charges is to charge polluters a fixed price for each unit of pollution. In this way, they are provided with an incentive to economize on the amount of pollution they produce.

Marketable permits represents the polar opposite of technology-based standards. Technology-based standards allow a firm to emit a specified amount, but do not allow a firm to trade this right to emit. Marketable permit schemes, on the other hand, define an overall level of emissions rights for all firms and then allow firms to trade them freely. Each permit enables the owner to emit a specified amount of pollution. In implementing this option the regulator must first specify a method for allocating permits to firms. Then, firms can choose abatement technologies based on the cost of buying permits to cover their excess emissions and the cost of pollution control. Like emissions fees, marketable permits have been advo-

---

[3] The words "fees" and "charges" will be used interchangeably. "User fees" refer to fees which are designed to cover some or all of the cost of an activity, but which do not necessarily provide appropriate incentives for efficient resource utilization. Example include water charges, sanitation fees, and airport ticket taxes.

cated by economists because of their desirable efficiency characteristics.

These four types of instruments provide a convenient taxonomy for discussing approaches that are frequently used in environmental policy. However, not all instruments easily fit into this categorization. For example, deposit-refund systems represent a mixture of taxes and subsidies. In some cases, instruments are completely overlooked. Notably absent from the preceding list is a discussion of liability rules and different types of enforcement mechanisms (Cohen and Rubin, 1985; Huber, 1987; Russell, Harrington and Vaughan, 1986). These mechanisms can be quite important in inducing firms to comply with regulations. Moreover, they are integrally linked to the type of approaches examined here. However, a detailed discussion of the use of enforcement mechanisms is beyond the scope of this work.

2.3.2. Instrument Choice and Efficiency

The relative merits of different instruments depends on the nature of information available to the regulator. In a world where the regulator has accurate information on individual firms costs, it is possible to use any of these instruments to design policies that yield outcomes that are highly cost-effective. In most applications, however, it is unreasonable to presume that the regulator has such knowledge. Consequently, schemes that require the regulator to dictate in detail the nature of the cleanup activity that firms should undertake will rarely be cost-effective. This is one of the primary criticisms of the standards approach as it has been implemented.

In the case of uniform standards, which is used in some applications to water pollution, it is usually possible to achieve significant cost savings by redistributing the cleanup burden so that firms for whom it is cheaper will abate more than firms with high abatement costs. Even in the case where standards are designed to approximate a least-cost solution, it is quite likely that the regulator will lack the information to identify the solution. In particular, one would expect that several industries possess proprietary information (i.e., not available to the regulator) on useful process modifications that would improve environmental quality. It would be desirable to employ an instrument that would induce industry to use such abatement options when they are cost-effective.

Another more serious flaw of the standards approach is that firms have no reason to abate more than the standard. In the most idyllic of worlds, where standards are treated as given, firms may have an incentive to search for lower cost alternatives that meet the standard; however, this will not always be the case since some standards are technology-based. If, instead of a standards approach, a pricing mechanism were used to reduce pollution, then, at least in theory, firms would have a continuous incentive to innovate—not only to find lower cost methods of achieving a given standard, but also to search for ways to reduce emissions.

While subsidies can provide firms with such a price signal, they are generally opposed by economists on the grounds that they often lead to inefficiencies. For example, if the cleanup of pollution is subsidized, it is possible that we could get too much pollution cleanup relative to other investments. In fact, pollution cleanup is subsidized in several ways. For example, some investments in pollution control equipment can qualify for tax-free financing and special tax credits. In addition, there are large direct federal subsidies in the form of grants for certain types of clean up activities, such as sewage treatment.

The theoretical case for or against subsidies is less clear and depends on how they are designed and the problem to which they are being applied. In theory, subsidies can be thought of as the mirror image of a tax. In some cases, they can be shown to provide an incentive for too many firms to enter a particular business, thus leading to inefficiencies (Page, 1973; Baumol and Oates, 1975). At the same time, however, approaches that rely on some form of subsidy may be helpful in identifying and properly disposing of hazardous waste (Hahn, 1988).

The two most highly touted economic approaches for meeting environmental objectives are effluent fees and marketable permits. While the two instruments share certain desirable characteristics, they differ dramatically in their application. Marketable permits are thought of as a quantity tool because the regulator chooses an upper bound on the allowable level of pollution. This is in contrast to emissions charges, which are a pricing tool.

The reason that economists support these approaches is that, in principle, they lead to outcomes which are cost-effective, even when the regulator has very little information on individual firm cost functions. Theory suggests that a polluter will abate up to the point where the marginal cost of abatement just equals the price of

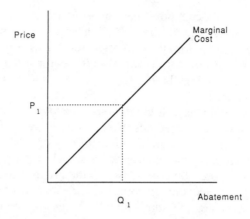

FIGURE 4   How Price and Quantity Mechanisms Work

pollution. If all firms are charged the same price for pollution, then marginal costs of abatement are equated across firms, and this result implies that the solution is cost-effective.[4] The normative properties of both these instruments have been studied in theory, and they are well understood in simple settings (Bohm and Russell, 1985; Montgomery, 1972).

Because marketable permits and emissions fees have desirable efficiency properties, they will be referred to as "incentive-based" instruments. The name is somewhat misleading since all instruments inevitably provide incentives. What these instruments provide are incentives for firms to achieve environmental targets using the fewest overall resources. Approaches that make use of marketable permits, but not emissions fees, will generally be referred to as "market-based" or "market-oriented" approaches.

The mechanism by which the incentive-based approaches work is suggested in Figure 4. Suppose an effluent fee is set at P_1. Even though the regulator may not know the actual marginal abatement cost schedule, suppose it is the one shown in the figure. The regulator observes that the associated level of abatement is Q_1. Assuming each firm minimizes costs, the firm's decision is relatively straightforward. A firm pays the tax when it is cheaper than abating

---

[4] The abatement cost schedule is typically assumed to be convex over the relevant region.

pollution, and it installs control equipment when it is less expensive than paying the tax. Thus, the tax serves as a type of equilibrating mechanism. Since the tax is the same for all firms, the continuous case shown in the figure dictates that the marginal cost of abatement will be equal to the tax for all firms.[5] An analogous situation holds for marketable permits. Suppose Q_1 property rights are issued to firms that allow firms to emit Q_1 tons of pollution per year. If the market for property rights is competitive, so firms can be treated as price takers in this market, the price that will emerge for a permit is P_1. Again, the result will be that overall costs are minimized since the marginal cost of abatement will equal the price of a permit for all firms.

These results on marketable permits and emission fees are deceptively simple. They focus on a single dimension of policy: cost-effectiveness. Moreover, they do so in a stylized manner. No mention has been made of the administrative costs of these systems or changes that might need to be made in monitoring and enforcement mechanisms. To illustrate the fragility of this paradigm, consider a simple application of monitoring and enforcement. Suppose firms could choose not to comply with the law and pay an expected fine of F. This possibility is illustrated in Figure 5. Figure 5 is identical to Figure 4, except for the addition of an expected fine of F, which is less than P_1. Now, consider the case of an effluent fee set at P_1. Suppose all firms are risk neutral so that they only care about their expected profits. Then the actual level of abatement will be Q_2 units. Note that the emissions fee no longer determines the level of abatement in this example. In general, the *lower* of the tax and the fine will determine the level of abatement (Hahn, 1982). The same type of result will obtain if Q_1 permits are issued. The price of a permit will fall to F, and there will be Q_2 units of abatement. It is instructive to compare the case where noncompliance is explicitly considered to the case where compliance is assumed. Note that the level of abatement decreases (e.g., from Q_1 to Q_2) when the expected fine is lower than the emissions fee. If the expected fine is above the emissions fee, then the level of abatement remains at Q_1, since the fee is now the binding constraint. The

---

[5] This efficiency result can be extended to other cases where technologies are discrete or continuous. The critical assumption relates to the convexity of the cost function.

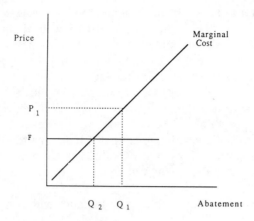

FIGURE 5 The Linkage between Emissions Fees, Marketable Permits and the
Structure of Enforcement

analysis for marketable permits is similar. The example highlights
the potential importance that the structure of enforcement can have
on the performance of the regulatory system.

Another factor confounding the simple argument about fees and
marketable permits is uncertainty over actual costs and benefits.
Weitzman (1974) has shown how the choice of a quantity or pricing
instrument will depend on the shape of the marginal benefit and
cost functions. In a different context, Roberts and Spence (1976)
illustrate how a mixed system will dominate a pure price or quantity
system when there is uncertainty over costs. Following a similar line
of reasoning, Hahn (1982) shows how mixed systems will generally
dominate when there is uncertainty over how firms will react in the
presence of imperfect monitoring and enforcement.

Taken together, these results on price and quantity instruments
suggest that the basic economic framework is a useful starting point
for examining the effect of different environmental policies. However, it needs to be applied with great care. Firm response to
different approaches can be dramatically affected by industry
structure (Borenstein, 1985; Hahn, 1984). Implementation issues
related to monitoring and enforcement will also affect the appropriate choice of instruments. Moreover, instruments can and should be
evaluated on a variety of dimensions.

2.3.3. Estimating Efficiency Gains

To gain a feel for the potential cost savings associated with the use of these systems, economists have performed a series of mathematical simulations linking cost and environmental quality. Typically, the cost of a system of uniform standards across sources is compared with an optimal system that could in theory be reached through the use of marketable permits or emission fees. The upshot of this research is that in many applications, the costs of achieving environmental quality goals could be significantly reduced if these instruments were used. For example, in a review of several studies examining the potential for marketable permits, Tietenberg found that potential control costs could be reduced by more than 90 percent in some cases (Tietenberg, 1985, pp. 43–44).

Some scholars have questioned the assumption that these instruments will perform well. For example, in the case of marketable permits, there could be serious problems with market domination if there were only a few firms in the market, or a single firm was able to corner the market (Hahn, 1984). However, even in such cases, a market may have significant advantages over a situation in which no trading is allowed across firms. One reason is that firms are assumed to engage in trading permits only if it is in their interest, and it will be in their interest to do so if there are cost savings involved. Thus, to the extent that firm motives are based on cost savings, gains from this approach may be substantial even in the presence of market power. However, if firms attempt to use the market in environmental property rights as a way of deterring entry into other markets, this may lead to significant problems. Fortunately, the ability of firms to pursue such a strategy may be limited in practice (Hahn and Noll, 1982a).

In addition to performing simulations to test the effectiveness of marketable permits and emissions charges, economists have also performed a series of controlled experiments that use human subjects to test how these ideas might work in practice. The idea behind these experiments is to provide subjects with similar incentives to the ones that firms might face under an emissions charge or marketable permit approach, and then examine how these approaches perform relative to some baseline. Work by Plott (1983) and Hahn (1983) reveals that implementation of these ideas in a laboratory setting leads to marked increases in efficiency levels over traditional forms of regulation, such as setting standards on a source by source basis.

In summary, the normative case for emissions charges and marketable permits has strong theoretical support. Moreover, simulations and laboratory experiments provide further support for the view that these approaches have the potential to make environmental policy more efficient.

2.4. The Politics of Instrument Choice

The design of instruments is not likely to be terribly useful if it does not take political realities into account. An instrument with wonderful efficiency properties, but which ignores political concerns is likely to remain a theoretical curiosity. Politics can be factored into the analysis using two distinct approaches drawn from positive political theory. The first approach is to consider how specific interest groups are likely to react to different kinds of policy proposals. The basic question each group can be expected to ask is, "What's in it for us?". Several different interest groups will be examined in the subsequent analysis. They include legislators, bureaucrats charged with developing and implementing regula- tions, industries that have to comply with regulations, environmen- talists concerned about cleaning up pollution, and academic groups interested in devising and promoting alternative approaches to regulation. The primary interest of legislators is frequently assumed to be reelection to the same office or a higher office, particularly in the Congress (Mayhew, 1974). Bureaucrats are less easily pinned down. Some scholars have argued that bureaucrats are interested in maximizing their discretionary resources and perks (Niskanen, 1971). In the context of the reforms examined in this paper, this theory is not sufficient to explain bureaucratic behavior. If there are any gains to bureaucrats from initiating incentive-based programs, they appear to come in the form of increased recognition as an entrepreneur who helped successfully initiate a regulatory reform (Hahn and Hester, 1986). Industrial and environmental lobbying groups often take strong positions on proposed environmental reforms. To the extent that one can generalize, it is probably fair to say that industry tends to support regulations that will result in reductions in direct costs. Environmentalist support for proposals

depends on the environmental quality impacts, and the environmental targets that are selected. Academics have a stake in trying
to get their ideas implemented.

To get a feel for how this analysis can be applied, consider a case
where an industry is trying to decide whether it wants a tax or
performance standards. Buchanan and Tullock (1975) have
analyzed this case in detail. Their basic approach can be gleaned
from a reinterpretation of Figure 4. Suppose that Figure 4 now
represents a choice for a single firm. The firm has two options. The
first is to have a performance standard set at Q_1, and the second is
to be subject to an emissions fee set at P_1. In both cases, the level of
abatement will remain the same. However, in the case of the tax,
the firm will not only have to pay for the abatement equipment, but
also have to pay the tax on its remaining emissions. Thus, based on
this simple analysis, it would appear to prefer the performance
standard. The primary difference between this positive analysis and
the normative analysis discussed above is that the positive analysis
focuses directly on distributional concerns and the immediate
interests of key economic and political actors.

There is a growing literature that attempts to grapple with the
question of why different instruments are actually selected in
different settings. Initially, scholars tried to identify the conditions
under which a quantity based instrument, such as marketable
permits or source specific standards, would be chosen over a fee.
Buchanan and Tullock (1975) argue that firms will prefer emission
standards to emission taxes because they result in higher profits.
Emission standards serve as a barrier to entry to new firms, thus
raising firm profits. Charges, on the other hand, do not preclude
entry by new firms, and also represent an additional cost to firms.
Their argument is based on the view that industry is able to exert
its preference for a particular instrument because it is more likely to
be well organized than consumers. Since this seminal article, several
authors have explored the instrument choice problem using this
basic framework (Coelho, 1976; Dewees, 1983; Yohe, 1976). The
basic insight of this work is that the argument that standards will be
preferred to taxes depends crucially on the precise nature of the
instruments being compared. In a different context, Campos (1987)
develops a model which views instrument choice as being controlled by legislators. Campos examines the motivation underlying

the choice of price supports or quotas in agriculture, and finds that the answer depends on both the demand for the commodity and the nature of the constituency support that the legislator attempts to nurture.

A somewhat more abstract approach is taken by Becker (1983) who assumes that groups compete for influence in an attempt to redistribute "the pie" to their benefit. Becker argues that governments will tend to choose mechanisms that are more efficient over those which are less efficient in redistributing revenues from less powerful to more powerful groups. One of the features of much existing environmental regulation is that it appears to result in high degrees of inefficiency. However, this does not necessarily refute Becker's theory since the inefficiencies may result from interest group pressures. If there is no more efficient means for redistributing revenues given interest group preferences, then Becker may be right. One problem with Becker's theory, however, is that it may not be testable in its current form because of the difficulty in specifying the influence functions. After presenting the various case studies, I will return to this theoretical result on efficient mechanisms and see how it squares with the facts.

This brief overview of different interest groups illustrates the diversity of perspectives that can influence proposed environmental reforms. It also illustrates how distributional issues are likely to affect attitudes towards different reform proposals. However, it ignores an important feature of any proposed reform, which is the context in which the reform is made. Context is important because it, too, can affect the viability of a reform. A key variable in understanding context is the institutional structure in which decisions are made. Institutional analysis represents a second approach for understanding the nature of proposed reforms. It can be used to analyze the political feasibility of different designs, as well as their likely results. In the case of federal environmental reform in the U.S., there are three key institutions: Congress, the courts, and the U.S. Environmental Protection Agency (EPA). While a detailed description of these institutions is beyond the scope of this book, the institutional context for proposed environmental reforms will be used in conjunction with the interest group approach to gain a deeper understanding of the politics of regulatory reform.

2.5. Summary

Large amounts of resources have been spent on improving environmental quality, especially in developed countries. The principal methods for regulating environmental quality have been through the setting of standards and the use of subsidies. Critics of current approaches, most notably economists, have noted that it may be possible to do more with less. These claims are largely based on beliefs that emerge from theoretical models. However, theory often diverges dramatically from practice. We now have a unique opportunity to examine the application of some incentive-based systems that have served as the rallying cry for the economics community. These policies will be evaluated using both the normative economics paradigm used in applied welfare economics and the positive interest-group paradigm used by political scientists. The normative theory will reveal how various reforms have performed. The positive theory will provide a deeper understanding of the potential and limitations of reform.

3. ECONOMIC PRESCRIPTIONS FOR ENVIRONMENTAL PROBLEMS: NOT EXACTLY WHAT THE DOCTOR ORDERED

3.1. Introduction

It is not easy to sit in an ivory tower and think of ways to help solve the world's environmental problems. As one who frequently engages in this exercise, I can attest to this fact. One of the dangers with ivory tower theorizing is that it is easy to lose sight of the actual set of problems which need to be solved, and the range of potential solutions. In my view, this loss of sight has become increasingly evident in the theoretical structure underlying environmental economics, which often emphasizes elegance at the expense of realism.

In this section, I will argue that both normative and positive theorizing could greatly benefit from a careful examination of the results of recent innovative approaches to environmental management. The particular set of policies examined here involve two tools that have received widespread support from the economics community—marketable permits and emission charges (Pigou,

1932; Dales, 1968; Kneese and Schultze, 1975). Both of these tools represent ways to induce businesses to search for lower cost methods of achieving environmental standards. Under highly restrictive conditions, it can be shown that both of these approaches share the desirable feature that any gains in environmental quality will be obtained at the lowest possible cost (Baumol and Oates, 1975).

The enthusiasm that many economists share for these tools is largely based on the belief that they have the potential to promote economic efficiency. Whether they do so in practice is another issue. Until the 1960's, these tools only existed on blackboards and in academic journals, as products of the fertile imagination of academics. Thus, it was only possible to speculate on how they might be applied in practice. However, recently countries have begun to explore using these tools as part of a broader strategy for managing environmental problems. This section chronicles the experience with both marketable permits and emission charges. It provides a selective analysis of a variety of applications in Europe and the United States. As we shall see, the actual use of these tools departs from the role which was initially conceived for them by economists.

The section has three objectives. The first is to provide a comprehensive review of the experience with these new approaches. The second is to identify important themes which emerge in the application of these tools. For example, does the implementation of these tools depart from the economists' prescriptions in systematic ways? The third objective is to assess how the introduction of these tools is shaped by broader political forces that are all-too-often ignored by economists. By gaining a deeper understanding of the political environment in which implementation occurs, we can begin to assess both the potential and limitations of these new economic approaches.

To set the stage for the analysis, part 3.2 provides a critical examination of selected policies in the U.S. and Europe. General patterns in implementation and performance are identified in part 3.3, with an eye towards bridging the gap between theory and practice. The implications of the performance of these mechanisms for improving system design are also addressed. Finally, part 3.4 summarizes the major conclusions and identifies areas for future research.

3.2. The Slip Twixt the Cup and the Lip

The formal results in the positive and normative theory are elegant. Unfortunately, they are not immediately applicable to many issues of instrument choice encountered in the real world. One reason is that they are overly simplistic. Virtually none of the systems examined below exhibits the purity of the instruments that are the subject of theoretical inquiry. This section provides an overview of some of the more important applications that have been studied by academics. The presentation highlights those instruments which show a marked resemblance to marketable permits or emission fees. A careful examination of the actual application of these mechanisms will suggest fruitful avenues for extending the bounds of existing theory.

Table I illustrates the wide range of applications of emissions charges and marketable permits in developed countries. Each

TABLE I
Charges and Marketable Permits: An Overview

Part 1: Charges

Water	Air	Solid waste	Hazardous waste	Noise	Products
Australia		Australia			
			Denmark		
		Finland			
France	France	France		France	France
Germany	Germany	Germany	Germany	Germany	Germany
Hungary					
Italy		Italy			
	Japan			Japan	
Netherlands	Netherlands	Netherlands	Netherlands	Netherlands	Netherlands
	Norway	Norway			Norway
		Sweden			Sweden
				Switzerland	
U.K.				U.K.	
U.S.	U.S.		U.S.		U.S.

Part 2: Marketable Permits

Water	Air	Solid waste	Hazardous waste	Noise	Products
	Germany				
U.S.	U.S.		U.S.		

Sources: Barde (1986), Boland (1986), Brown (1984a), Brown (1984b), Brown and Bressers (1986), Hahn (1982), Hahn (1987a), Liroff (1986), Novotny (1986), Sprenger (1986), U.S. Congressional Budget Office (1985).

column of the table corresponds to a different application. The table is partitioned into two parts. The first part shows the countries that are using various charge systems. Charge systems are interpreted broadly here to include both user charges and emissions charges. User charges generally include charges for treating or storing waste, such as garbage. Emissions charges generally apply to charges on specific pollutants that are discharged into the environment.[6] The second part of the table shows countries that have adopted some form of marketable permits. The table was constructed on the basis of available information. It is not meant to be exhaustive. If a country does not appear in a particular category, this does not necessarily imply that it does not have a program in that area; only that the author is unaware of one. Moreover, if a country is designated as using a particular program, it may do so on a small or a large scale, depending on the application. For example, the U.S. makes only nominal use of air pollution charges in selected areas. These charges have a relatively small effect on revenues and no discernible effect on pollution levels.

An inspection of the table reveals several interesting points. First, the use of charges is widespread in the developed world, in the sense that several countries have experimented with their use. Second, the application of marketable permits to environmental problems appears to be quite limited. There are only two major applications of this concept, both found in the United States. Outside of the U.S., markets do not appear to have received widespread use. Indeed, there is only one application outside of the United States that appears to be significant (Sprenger, 1986, pp. 18–22).[7] The final point to be gleaned from the table is that applications of these mechanisms span a wide array of environmental problems.

Table I provides an overview of existing applications, but it does not offer information on the implementation and performance of

---

[6] Barde (1986, p. 13) notes that the distinction between effluent charges and user fees is not always very clear. For example in the case of water, user fees and effluent charges are closely linked.

[7] Opschoor (1986, p. 17) suggests that limited forms of trading are used in the Netherlands, which are similar to the "bubbles" used in the U.S. (see discussion on page 33.) However, no data were provided on the magnitude of trading that has taken place. Lidgren (1986, p. 8) also notes that there have been some minor experiments with the use of bubbles in Sweden.

various instruments. To gain a better understanding of how instruments work, it is necessary to systematically explore different applications. The following discussion presents a wide array of applications in different countries. The purpose of the analysis is to uncover important trends in the implementation of these mechanisms. Following the organization in Table I, selected charge mechanisms in different countries are examined first. This is followed by a discussion of the few existing marketable permit approaches in the United States.

3.2.1. Charges in Practice

Applications of charge systems are best understood in terms of the broader regulatory context in which they are implemented. The cases reviewed here provide background on the regulatory structure in which these reforms are initiated. At the same time, they attempt to summarize salient features of these instruments that will yield insights into the political and economic foundations of environmental policy.

Charge systems in four countries will be explored in some detail. Examples are drawn from France, Germany, the Netherlands, and the United States.[8] Particular systems were selected because they were thought to be significant either in their scope, their effect on revenues, or their impact on the cost-effectiveness of environmental regulation. While the focus is on water effluent charges, a variety of systems are briefly mentioned at the end of this section that cover other applications.[9]

3.2.1.1. France

The French have had a system of effluent charges on water pollutants in place since 1969 (Bower et al., 1981).[10] The system is primarily designed to raise revenues, which are then used to help maintain or improve water quality. For purposes of managing water, France is divided into a series of basins, which are further divided into zones. Charges are levied by the basin agencies, and set

---

[8] Throughout this chapter, Germany refers to the Federal Republic of Germany.

[9] The OECD is in the process of collecting information on a variety of incentive systems currently in place in OECD countries. See generally Barde (1986) for a discussion of the status of this work.

[10] As used here, effluent charges will be used to denote emissions charges related to discharges into waterways.

through a complex procedure in which the basin agency particip-
ates. The basin agency also receives the payment of the charges.
Revenues from the charges are then redistributed to the dischargers through the use of grants, subsidized loans, and rewards for
superior performance (Bower et al., 1981 p. 137). In addition, they
are used to help support the basin agencies. Charges cover a wide
variety of pollutants, including suspended solids, biological oxygen
demand (BOD), chemical oxygen demand (COD), and selected
toxic chemicals. Charges, though widespread, are relatively low.
Moreover, charges are rarely based on actual performance. Rather,
they are based on the expected level of discharge by various
industries. There is no explicit connection between the charge paid
by a given discharger and the subsidy received (Bower et al., 1981,
p. 126). However, charges are generally earmarked for use in
promoting environmental quality in areas related to the specific
charge.

While the majority of charges are based on average as opposed to
individual performance, some charges are based on actual behavior.
If the activity involves a treatment plant, performance incentives
are related to actual behavior. This mechanism helps promote more
efficient operation of treatment plants (Bower et al., 1981, p. 130).
Firms can also ask for individual monitoring of their behavior if
they believe, for example, their performance is above average.
Moreover, if firms are thought to be exceeding some maximum
allowable value dictated by their permit, penalties can be imposed.
Thus, incentives related to individual behavior do exist; however,
they are rather blunt in that they do not provide a continuous
incentive to search for lower cost options of abating pollution.

There is little quantitative data that can be used to measure the
effect of the effluent charges; however, there is a general impression
that the system has worked in the sense that water quality has
improved (Bower et al., 1981, pp. 23–24). While the system may
help result in environmental improvements, the charges do little in
and of themselves to promote cost-effective environmental improvement. This is because they are relatively low in magnitude and
are based on average performance rather than individual performance. Moreover, the charges are generally based on water intake.
Therefore, the charges do not provide any incentive for appropriate
use or discharge. The basic mechanism by which these charges

improve environmental quality is through judicious earmarking of the revenues for pollution abatement activities.

In evaluating the charge system, it is important to understand that it is a major, but by no means dominant, part of the French system for managing water quality. Indeed, in terms of total revenues, a sewage tax levied on households and commercial enterprises is larger in magnitude (Bower *et al.*, 1981, p. 142). Moreover, this tax is assessed on the basis of actual volumes of water used.

Like most other charge systems, the charge system in France is based on a system of water quality permits, which places constraints on the type and quantity of effluent a firm may discharge. These permits are required for sources discharging more than some specified quantity (Bower *et al.*, 1981, p. 130).

Charges appear to be accepted as a way of doing business in France now. They provide a significant source of water quality control. One of the keys to their initial success appears to have been the gradual introduction and raising of charges. Charges started at a very low level and were gradually raised to current levels (Bower *et al.*, 1981, p. 22). Moreover, the set of pollutants on which charges are levied has expanded considerably since the initial inception of the charge program.[11]

3.2.1.2. Germany

The German system of effluent charges is very similar to the French system. Effluent charges cover a wide range of pollutants including settleable solids, BOD, COD, cadmium, and mercury (Brown and Johnson, 1984, pp. 934, 939). The charges are used to cover administrative expenses for water quality management and to subsidize projects that improve water quality (Brown and Johnson, 1984, p. 945).

In 1981, a system of nationwide effluent charges was introduced (Bower *et al.*, 1981, p. 226). Charges have existed in selected areas of Germany, such as the Genossenschaften, for decades (Bower *et al.*, 1981, p. 229). Management of water quality is delegated to local areas. The federal government provided the basic framework in its 1976 Federal Water Act and Effluent Charge Law (Brown and

---

[11] For example, Brown, 1984a, p. 114, notes that charges for nitrogen and phosphorous were added in 1982.

Johnson, 1984, p. 930). States generally set water quality standards. Local areas can choose to implement stricter standards if they desire.

Primary responsibility for implementation and enforcement of the system is given to the "Lander", which are the equivalent of states (Brown and Johnson, 1984, p. 930). Charges are implemented in conjunction with water quality permits. The bills that industry and municipalities pay are generally based on expected volume and concentration (Brown and Johnson, 1984, p. 934). If a firm meets its actual standard, it is given a 50% discount. However, if firms exceed maximum allowable volumes or concentration, the charge can be raised (Brown and Johnson, 1984, p. 934). Charges vary by industry type as well as across municipalities. Charges to municipalities depend on several variables, including size of the municipality, desired level of treatment, and the age of equipment (Brown and Johnson, 1984, p. 938).

Like the French system, it is difficult to measure the effect of the German effluent charge system. Initially, charges were opposed by industry (Brown and Johnson, 1984, p. 932).[12] However, there is a general perception that the current system is helping to improve water quality. Unfortunately, no direct data on the impact of charges were found. At a minimum, the charge system serves as an effective subsidy for water quality management projects.

3.2.1.3. Netherlands

The Netherlands has had a system of effluent charges in place since 1969 (Brown and Bressers, 1986, p. 4). It is one of the oldest and best administered charge systems, and the charges placed on effluent streams are among the highest. In 1983, the effluent charge per person was $17 in the Netherlands; $6 in Germans and about $2 in France (Brown and Bressers, 1986, p. 5). Because of the comparatively high level of charges found in the Netherlands, this is a logical place to examine whether charges are having a discernible effect on the level of pollution. Bressers (1983), using a multiple regression approach, argues that charges have made a significant difference in the levels of BOD and heavy metals. This evidence is

---

[12] After losing the initial battle, industry focused on how charges would be determined and their effective date of implementation (Brown and Johnson, 1984, p. 932).

also buttressed by surveys of industrial polluters and water board officials which indicate that charges had a significant impact on firm behavior (Brown and Bressers, 1986, pp. 12–13). This analysis is one of the few existing empirical investigations of the effect of effluent charges on resulting pollution.

The purpose of the charge system in the Netherlands is to raise revenue, which is used to finance projects that will improve water quality (Brown and Bressers, 1986, p. 4). Like its counterparts in France and Germany, the approach to managing water quality uses both permits and effluent charges for meeting ambient standards (Brown and Bressers, 1986, p. 2). Permits tend to be uniform across similar dischargers. The system is designed to ensure that water quality will remain the same or get better (Brown and Bressers, 1986, p. 2). Charges are administered both on expected and actual levels of discharge. Actual levels of discharge are monitored for larger polluters, while small polluters often pay fixed fees unrelated to actual discharge (Bressers, 1983, p. 10).

Several pollution measures are used including COD, heavy metals and nitrogen. One set of charges is based on a metric defined as "population equivalent", which represents a weighted sum of COD and nitrogen.[13] The charge system is administered by Water Boards and the Ministry of Transport and Public Works (Brown and Bressers, 1986, p. 3). Charges are imposed both on volume and concentration. Revenues from charges on outputs are about fifteen times the revenue from volume charges (Brown and Bressers, 1986, p. 9). There is some variation in charges across jurisdictions. For example, in 1983, there was a threefold difference between the lowest and highest charge (Brown and Bressers, 1986, p. 5).

Charges have exhibited a slow, but steady increase since their inception (Brown and Bressers, 1986, p. 5). This increase in charges is correlated with declining levels of pollutants. Effluent discharge declined from 40 population equivalents in 1969 to 15.3 population equivalents in 1980, and it was projected to decline to 4.4 population equivalents in 1985 (Brown and Bressers, 1986, p. 10). Thus, over 15 years, this measure of pollution declined on the order of 90%.

---

[13] Formally, a population equivalent is given by the formula, load (population equivalent) = $(cod + 4.57N)/180$ (Brown and Bressers, 1986, p. 2).

As in Germany, there was initial opposition from industry to the use of charges. Brown and Bressers (1986, p. 4) also note opposition from environmentalists, who tend to distrust market-like mechanisms. Nonetheless, charges have enjoyed widespread acceptance in a variety of arenas in the Netherlands. As in Germany and France, charges started at comparatively low levels and then were gradually increased.

One final interesting feature of the charge system in the Netherlands relates to the differential treatment of new and old plants. In general, newer plants face more stringent regulation than older plants (Brown and Bressers, 1986, p. 10). As we shall see, this is also a dominant theme in American regulation.

3.2.1.4. United States

The United States has a modest system of user charges levied by utilities that process wastewater. Federal environmental regulations issued by the Environmental Protection Agency have encouraged the use of these charges. The charges cover a variety of targets including BOD, COD, and total suspended solids. They are based on both volume and strength, and vary across wastewater utilities. In some cases, charges are based on actual discharges and in others, a rule of thumb, related to average behavior.[14] In some cases, water quality permits require that pretreatment be performed. Local areas have some latitude in setting the charges provided they conform to federal guidelines. In all cases, charges are added on to the existing regulatory system, which relies heavily on permits and standards.

Both industry and consumers are required to pay the charges. The primary purpose for the charges is to raise revenues to help meet the revenue requirements of the wastewater utilities, which are heavily subsidized by the federal government. The direct environmental and economic impact of these charges is apparently small (Boland, 1986, p. 12). They primarily serve as a mechanism to help defray the costs of the treatment plants. Thus, the charges used in the United States are similar in spirit to the German and French

---

[14] For example, see Boland, 1986, p. 12. Dischargers using more than 25,000 gallons per day are required to be charged on their actual use.

system already described. However, their size appears to be smaller, and the application of the revenues is more limited.

3.2.1.5. Other Fee Based Systems

There are a variety of other fee based systems which have not been included in this discussion. Brown (1984a) did an analysis of incentive-based systems to control hazardous wastes in Europe and found that a number of countries had adopted systems, some of which had a marked economic effect. The general trend was to use either a waste-end tax or a tax on feedstocks. Companies and government officials were interviewed to ascertain the effects of these approaches. In line with economic theory, charges were found to induce firms to increase expenditures on reducing discharges. Waste reduction was achieved through a variety of techniques including reprocessing of materials, treatment, and input and output substitution (Brown, 1984a, p. 3). Firms also devoted greater attention to separating waste streams because prices for disposal often varied by the type of waste stream.

One of the systems reviewed was a charge system in Denmark used for the disposal of hazardous wastes. Charges vary by volume and the quantity of the wastes. Based on interviews with key participants, Brown found that the charge system had the expected effects. In particular, it encouraged the separation of waste streams so that expensive charges would be avoided. It also encouraged recovery of waste streams, for example, through techniques such as reprocessing and increased recovery (Brown, 1984a, p. 12).

Another effective application of fees for reducing hazardous wastes arises in Bavaria, Germany, where a detailed charge system has been in place for fifteen years (Brown, 1984a, p. 19). On the basis of prices for treatment and disposal, companies decided how much treatment they want to undertake individually and how much will be performed at jointly owned treatment centers. Charges generally increase with the difficulty of treatment and the concentration of the waste stream. They decrease as the useful caloric content of the waste increases. Companies were found to be quite responsive to this system. For example, as charges increased from $22 per ton in 1974 to $82 per ton in 1980, BMW decreased its quantity of waste delivered from 6,300 tons to 3,800 tons (Brown, 1984a, p. 25).

The United States has a diverse range of taxes imposed on hazardous waste streams. Several states have waste-end taxes in place. Charges exhibit a wide degree of variation across states. For example, in 1984, charges were $.14/tonne in Wisconsin and $70.40/tonne in Minnesota (U.S. CBO, 1985, p. 82). Charges for disposal at landfills also vary widely. The effect of these different charges is very difficult to estimate because of the difficulty in obtaining the necessary data on the quantity and quality of waste streams, as well as the economic variables.

In addition to taxing waste streams, several countries impose taxes on products that are related to waste streams. The United States, for example, has a system of taxes on petroleum and chemical feedstocks. Between 1981 and 1984, these taxes generated over $800 million in revenues, which will be used to help finance hazardous waste cleanup under Superfund (Hahn, 1988, p. 12). While the amount of revenue raised seems large, the tax is small relative to the price of the inputs, and appears to have had little or no effect on firm behavior.

Several countries place taxes on oil and oil-related produces based on their potential for environmental harm. Notable examples include Germany, France and the Netherlands. In 1969, Germany passed the Used Oil Statute, which calls for a system of charges and subsidies on waste oil (Brown, 1984a, p. 28). Firms receive a subsidy if they accept and dispose of waste oil properly. Subsidies are based on the costs of disposal for an average firm. The subsidies are financed by charges that are levied on lubrication oils. Brown argues that the tax/subsidy scheme has been a success, pointing to the fact that 98% of the waste is being treated in a safe manner (Brown, 1984a, p. 33). In addition, firms disposing of the waste have an incentive to search for least cost solutions, since their compensation is based on the quantity of oil they take in and not the precise technique they use for treating the waste oil.

Most of the fee-based systems discussed up to this point have involved water and hazardous waste. There are some fee systems related to air pollution, but they are less prominent. For example, Sweden has a gasoline tax that increases with the lead and sulfur concentration of the gasoline (Brown, 1984b). The United States imposes nominal fees on various emissions to help pay for some of the administrative costs associated with regulation. Both Los Angeles and Wisconsin have nominal fees associated with air emissions (Hahn and Noll, 1982a; Hahn, 1987b).

The preceding analysis reveals that there are a wide array of fee-based systems in place designed to promote environmental quality. The structure of these systems is remarkably similar in many ways. Almost all of the fees are used to subsidize some aspect of environmental quality. In a few cases, the fees were shown to have a marked effect on firm behavior; however, in the overwhelming majority of cases studied, the direct economic effect of fees appears to have been small. This is largely due to the fact that fees have been set at comparatively low levels, and that fees are often only tangentially linked to the actual behavior of individual firms.

3.2.2. Marketable Permits

In comparison with fees, marketable permits have not received widespread use. Indeed, there appears to be only four existing environmental applications, three of which have been implemented in the United States. One of these applications involves the control of BOD in a limited river area. The other three are related to air pollution. One involves the trading of emissions rights of various pollutants regulated under the Clean Air Act; a second involves trading of lead used in gasoline; a third involves air pollution trading in Germany and will not be addressed here because of limited information (see Sprenger, 1986). As we shall see, the performance of these programs exhibits dramatic differences, which can be traced back to the rules used to implement the different mechanisms.

3.2.2.1. Wisconsin Fox River Water Permits

In 1981, the state of Wisconsin implemented an innovative program aimed at controlling BOD on a part of the Fox River (Novotny, 1986, p. 11). The program was designed to allow for the limited trading of marketable discharge permits. The primary objective was to allow firms greater flexibility in abatement options while still maintaining environmental quality. The program is administered by the state of Wisconsin in accord with the Federal Water Pollution Control Act. Firms are issued five year permits that define their wasteload allocations. This allocation defines the initial holding of permits for each firm. Allowable discharges to the river vary over the year as a function of both stream flow and temperature.

Early studies estimated that substantial savings, on the order of $7 million per year, could result after implementing this trading system (O'Neil, 1983, p. 225). However, actual cost savings have been minimal. In the six years that the program has been in existence, there has been only one trade. This trade took place between a paper company and a municipal treatment center (Patterson, 1987). Given the initial fanfare about this system, its performance to date has been disappointing.

A closer look at the nature of the market and the rules for trading reveals that the result should not have been totally unexpected. The regulations are aimed at two types of dischargers: pulp and paper plants and municipal waste treatment plants. Approximately 2/3 of the sources are pulp and paper mills, and the remaining 1/3 are municipal treatment plants (Patterson, 1987). David and Joeres (1983) note that the pulp and paper plants have an oligopolistic structure, and thus may not behave as competitive firms in the permit market. Moreover, it is difficult to know how the municipal utilities will perform under this set of rules, since they are subject to public utility regulation (Hahn and Noll, 1983). Trading is also limited by location. There are two points on the river where dissolved oxygen concentrations are critical, and firms are divided into "clusters" so that trading will not increase BOD at either of these points. There are only about 6 or 7 firms in each cluster (Patterson, 1987). Consequently, markets for wasteload allocations may be quite thin.

Novotny (1986) has argued that there are several restrictions on transfers that have a negative impact on potential trading. Any transaction between firms requires modifying or reissuing permits. This takes at least 175 days, and sometimes longer. Transfers must be for, at least, a year; however the life of the permit is only five years. Moreover, parties must waive any rights to the permit after it expires, and it is unclear how the new allocation will be affected by trading. This creates great uncertainty over the future value of the property right. Added to the problems created by these rules are the restrictions on eligibility for trades. Firms are required to justify the "need" for permits. This effectively limits transfers to new dischargers, plants that are expanding, and treatment plants that cannot meet the requirements despite their best efforts. Trades that only reduce operating costs are not allowed. With all the uncertainty and high transactions costs, it is not surprising that trading has gotten off to a very slow start.

While the marketable permit system for the Fox River was being hailed as a success by economists, the paper mills did not enthusiastically support the idea (Novotny, 1986, p. 15). Nor have the mills chosen to explore this option once it has been implemented. The approach is best viewed as a very limited form of permit trading. Indeed, by almost any measure, these transferable permits represent a minor part of the regulatory structure. The mechanism builds on a large regulatory infrastructure where permits specifying treatment and operating rules lie at the center. The new marketable permits approach retains many features of the existing standards-based approach. The initial wasteload allocations are based on the *status quo,* calling for equal percentage reductions from specified limits. This "grandfathering" approach has a great deal of political appeal for existing firms. New firms must continue to meet more stringent requirements than old firms, and firms must meet specified technological standards before trading is allowed. In this regard, the change is best viewed as incremental.

3.2.2.2. Emissions Trading
By far, the most significant and far-reaching program in the United States is the emissions trading policy.[15] Started over a decade ago, the policy attempts to provide greater flexibility to firms charged with controlling air pollutant emissions. Pollutants covered under the policy include volatile organic compounds, carbon monoxide, sulfur dioxide, particulates, and nitrogen oxides (Hahn and Hester, 1986, p. 31). Because the program represents a radical departure in the approach to pollution regulation, it has come under close scrutiny by a variety of interest groups. Environmentalists have been particularly critical of this reform. These criticisms notwithstanding, the EPA Administrator characterizes the program as

---

[15] This analysis of the emissions trading and lead trading programs draws heavily on joint work with Gordon Hester.

"one of EPA's most impressive accomplishments," (Thomas, 1986).

The Clean Air Act provides the authority for the regulation of air pollution in the U.S. Two features of the Act have proved particularly significant for emissions trading. First, the Act specifies that every significant source of air emissions will be regulated individually, and that different classes of sources will have different standards applied to them. This has led to large variations in marginal costs of abatement across sources. The second significant feature of the Clean Air Act affecting emissions trading is the provision that federal agencies determine the standards and structure of the regulatory system, but the states implement it. States often did not have the capabilities or resources to develop coherent plans for implementing environmental strategies. Many emissions inventories and projections were highly inaccurate. This led to a great deal of controversy over how to define the nature of property rights that would serve as the cornerstone of the emissions trading program. The emission limits contained in operating permits became the basis for the allocation of property rights for emissions trading. These property rights are referred to as "emission reduction credits."

There are now four distinct elements of emissions trading. Netting, the first program element, was introduced in 1974. Netting allows a firm that creates a new emission source in a plant to avoid the stringent emission limits that would normally apply by reducing emissions from another source in the plant. Thus, net emissions from the plant do not increase significantly. (However, a small increase in net emissions may result in some cases.) A firm using netting is only allowed to obtain the necessary emission credits from its own sources. This is called *internal trading* because the transaction involves only one firm. Netting is always subject to approval at the state level, not the federal.

Offsets, the second element of emissions trading, are used by new emission sources in "non-attainment areas."[16] The Clean Air Act specified that no new emission sources would be allowed in non-attainment areas after the original 1975 deadlines for meeting

---

[16] A non-attainment area is a region which has not met a specified ambient standard.

air quality standards passed. Concern that this prohibition would stifle economic growth in these areas prompted EPA to institute the offset rule. This rule specified that new sources would be allowed to locate in non-attainment areas, but only if they "offset" their new emissions by reducing emissions from existing sources by even larger amounts. The offsets could be obtained through internal trading, just as with netting. However, they could also be obtained from other firms' sources, which is called *external trading*. Like netting, offsets are subject only to state approval.

Bubbles, though apparently considered by EPA to be the centerpiece of emissions trading, were not allowed until 1979. The name derives from the placing of an imaginary bubble over a plant, with all emissions exiting at a single point from the bubble. A bubble allows a firm to sum the emission limits from individual sources of a pollutant in a plant, and to adjust the levels of control applied to different sources so long as this aggregate limit is not exceeded. While the trading concept for bubbles is similar to that for netting and offsets, bubbles apply to existing sources. Initially, every bubble has to be approved at the federal level as an amendment to a state's implementation plan. In 1981, EPA approved a "generic rule" for bubbles in New Jersey which allowed the state to give final approval for bubbles. Since then, several other states have followed suit.

Banking, the fourth element of emissions trading, was developed in conjunction with the bubble policy. Banking allows firms to save emission reductions above and beyond permit requirements for future use in emissions trading. While EPA action was initially required to allow banking, the development of banking rules and the administration of banking programs has been left to the states.

The performance of emissions trading can be measured in several ways. A summary evaluation which assesses the impact of the program on abatement costs and environmental quality is provided in Table II. For each emissions trading activity, an estimate of cost savings, environmental quality effect, and the number of "trades" is given. In each case, the estimates are for the entire life of the program. As can be seen from the table, the level of activity under various programs varies dramatically. More netting transactions have taken place than any other type, but all of these have necessarily been internal. The wide range placed on this estimate,

TABLE II
Summary of Emissions Trading Activity

Activity	Estimated number of internal transactions	Estimated number of external transactions	Estimated cost savings (millions)	Environmental quality impact
Netting	5,000 to 12,000	None	$25 to $300 in permitting costs; $500 to $12,000 in emission control costs	Insignificant in individual cases; probably insignificant in aggregate
Offsets	1800	200	See text	Probably insignificant
Bubbles: federally approved	40	2	$300	Insignificant
state approved	89	0	$135	Insignificant
Banking	<100	<20	Small	Insignificant

Source: Hahn and Hester (1986)

5,000 to 12,000, reflects the uncertainty about the precise level of this activity. An estimated 2000 offset transactions have taken place, of which only 10% have been external. Fewer than 150 bubbles have been approved. Of these, almost twice as many have been approved by states under generic rules than have been approved at the federal level, and only two are known to have involved external trades. For banking, the figures listed are for the number of times firms have withdrawn banked emission credits for sale or use. While no estimates of the exact numbers of such transactions can be made, upper bound estimates of 100 for internal trades and 20 for external trades indicate the fact that there has been relatively little activity in this area.

Cost savings for both netting and bubbles are substantial.[17] Netting is estimated to have resulted in the most cost savings, with a total of between $525 million and over $12 billion from both permitting and emissions control cost savings. The wide range of this estimate reflects the uncertainty that results from the fact that little information has been collected on netting. Offsets result in no direct cost savings in the sense that the use of offset does not allow a firm to avoid any emission limits. However, firms probably receive substantial benefits from the use of offsets because they are allowed to locate new or modified major emission sources in non-attainment areas, something which they would not be able to do without the use of offsets. The fact that firms are willing to go to the expense of obtaining offsets indicates that they derive some net gain from doing so, but the extent of this gain has not been estimated. Federally approved bubbles have resulted in savings estimated at $300 million, while state bubbles have resulted in an estimated $135 million in cost savings.[18] As these figures indicate, average savings from federally approved bubbles are higher than those for state approved bubbles. Average savings from bubbles are higher than those from netting, which reflects the fact that bubble savings may be derived from several emissions sources in a single transaction, while netting usually involves cost savings at a single source. Finally, the cost savings from the use of banking cannot be

---

[17] For a detailed discussion of the derivation of these figures, see Hahn and Hester (1986), pp. 43–44 and 50–52.

[18] These estimates include savings for bubbles under review as well as those already approved.

estimated, but is necessarily small given the small number of banking transactions that have occurred.

The effects that the various program elements have had on environmental qualtiy are, on the whole, insignificant. While there have been some small emissions increases from individual sources involved in netting transactions, the overall effect has been inconsequential. For offsets and bubbles, aggregate effects are also thought to be insignificant, although they may be slightly positive. Some transactions may have had an adverse impact on the environment due to the use of reductions in allowable (as opposed to actual) emissions, but their aggregate impact on individual areas is thought to be inconsequential. Banking has probably had a very slight positive effect, since banked credits represent emission reductions that have not been used to offset emission increases. However, due to the fact that there has been little banking activity, the total effect of banking is necessarily very small.

The performance evaluation of emissions trading activities reveals a mixed bag of accomplishments and disappointments. The program has clearly afforded many firms flexibility in meeting emission limits, and this flexibility has resulted in significant aggregate cost savings—in the billions of dollars. However, these cost savings have been realized almost entirely from internal trading. They fall far short of the potential savings that could be realized if there were more external trading. The scorecard on environmental effects is more difficult to estimate. However, the best available information indicates that the program has led to little or no change in the level of emissions.

The evolution of emissions trading can be best understood in terms of a struggle over the nature and distribution of property rights (Hahn and Hester, 1986). It involves not only concerns over measurable outputs such as costs and environmental quality, but also underlying values. Emissions trading can be seen as a strategy by regulators to provide industry with increased flexibility while offering environmentalists continuing progress toward environmental quality goals. Meeting these two objectives requires a careful balancing act. To provide industry with greater flexibility, EPA has attempted to define a set of property rights that places few restrictions on their use. However, at the same time, EPA had to be sensitive to the concerns of environmentalists regarding the definition of property rights. The conflicting interests of these two groups have led regulators to create a set of policies that are specifically designed to deemphasize the explicit nature of the property right. The high transactions costs associated with external trading have induced firms to eschew this option in favor of internal trading or no trading at all.

Like the preceding example of the Fox River, emissions trading is best viewed as an incremental departure from the existing approach. Property rights were grandfathered. Most trading has been internal, and the structure of the Clean Air Act, including its requirements that new sources be controlled more stringently, was largely left intact.

3.2.2.3. Lead Trading

Lead trading stands in stark contrast to the preceding two marketable permit approaches. It, by far comes the closest to an economist's ideal of a freely functioning market.[19] The purpose of the lead trading program is to allow gasoline refiners greater flexibility during a period when the amount of lead in gasoline was being significantly reduced. The program was designed, in part, to help ease the transition for small refiners. Decreasing the lead content in gasoline meant that most small refiners would need to install new equipment for producing gasoline in order to meet reduced lead content standards while maintaining octane ratings.

Unlike many other programs, the lead trading program was scheduled to have a fixed life from the outset. Inter-refinery trading of lead credits was permitted in 1982. Banking of lead credits was initiated in 1985. The program was terminated at the end of 1987.

Credits were allocated on the basis of the current standard and actual production levels. For example, in 1982, the lead standard was 1.1 grams/gallon for large refiners. Suppose a firm produced 10,000,000 gallons/year containing an average lead content of 1.0 grams/gallon. It would then have created a credit of $10,000,000 \times (1.1-1.0) = 1,000,000$ grams, which it could sell to other refiners.[20]

---

[19] For a more detailed discussion of the performance of lead trading, see Hahn and Hester (1987a).

[20] In practice, rights were allocated by calendar quarter, See 47 *Fed. Reg.* 49326-49327 for more details.

The standard has been reduced over time from 1.1 grams/gallon for large refiners in 1982 to 0.1 grams/gallon for all refiners in 1986.[21] Initially, the period for trading was defined in terms of quarters or three-month intervals. No banking of credits was allowed. Rights created in a quarter had to be used or traded in that quarter; otherwise they had no value. Three years after initiating the program, limited banking was allowed, which allowed firms to carry over rights to subsequent quarters. Banking has been used extensively by firms since its initiation. In comparison to other trading programs, the lead program makes little use of the existing permitting structure. This is because the permitting process plays a less direct role in regulating the use of lead in gasoline. Nonetheless, the lead trading program does build on the existing regulatory structure. The mechanisms used to monitor the lead content in gasoline under the old regulatory structure are also used in helping to monitor lead trading.

The program is notable for its lack of discrimination among different sources, such as new and old sources. It is also notable for its rules regarding the creation of credits. Lead credits are created on the basis of existing standards. A firm does not gain any extra credits for being a large producer of leaded gasoline in the past. Nor is it penalized for being a small producer.[22] The creation of lead credits is based solely on current production levels and average lead content. To the extent that current production levels are correlated with past production levels, the system does acknowledge the existing distribution of property rights. However, this linkage is less explicit than those made in other trading programs.

The lead credit allocation rule creates some interesting incentives. To see this, first consider a world in which the standard for all gasoline is 1.1 grams/gallon and rights are not tradable. Then, introduce trading with property rights defined on the basis of current production. The demand for gasoline is presumed to remain constant before and after trading. However, the supply curve would shift downward, due to the more efficient production that would be achieved through trading and the *increased* use of lead. Taken by itself, this allocation rule tends to increase both gasoline output and

---

[21] 47 *Fed. Reg.* 49326, and 50 *Fed. Reg.* 9386.

[22] This is not to suggest that large and small producers were treated alike in all respects. In general, the standards for small refiners were less stringent (47 *Fed. Reg.* 49326).

lead output. The degree to which these outputs would increase depends on the elasticity of demand for gasoline and the specific shift in the supply curve.[23]

It may seem curious that the implementers of the lead trading rule created a system which would theoretically result in the increased use of lead in the short term. One of the ostensible reasons for taking this approach is that it would induce producers to "compete away the rents" from the lead credits (Mannix, 1987). This, of course, will depend on the nature of the shift in the supply curve for gasoline. If the supply curve becomes "flatter" as it shifts downward, rents will be competed away. While this rule may have caused a short-term increase in lead, over the medium and longer term the ratcheting down of the standard resulted in sharp decreases in lead use.

The success of the lead trading program is difficult to measure directly. It appears to have had very little impact on environmental quality. This is because the amount of lead in gasoline is routinely reported by refiners and is easily monitored. The effect the program has had on refinery costs is not readily available. In proposing the rule for banking of lead rights, EPA estimated that resulting savings to refiners would be approximately $228 million (U.S. EPA, 1985a). Since banking activity has been somewhat higher than anticipated by EPA, it is likely that actual cost savings will exceed this amount. No specific estimate of the actual cost savings resulting from lead trading have been made by EPA.

The level of trading activity has been high, far surpassing levels observed in other environmental markets. In 1985, over half of the refineries participated in trading. Approximately 15% of the total lead rights used were traded. Approximately 35% of available lead rights were banked for future use or trading (U.S. EPA, 1985b, 1986).[24]

---

[23] The alternative of grandfathering lead rights on the basis of past production would also have resulted in higher gasoline production than the case of no trading, but without increasing the aggregate amount of lead. Compared to the allocation rule which EPA adopted, the grandfathering rule results in lower output levels of lead and gasoline. However, for the two cases in which trading is allowed, prices will be lower and gasoline output higher than the case of no trading.

[24] In contrast, even in the most vigorous markets for emissions trading, such as Los Angeles, less than 1% of the total rights associated with any given pollutant were traded (Hahn and Hester, 1987).

From the standpoint of creating a workable regulatory mechanism that induces cost savings, the lead market has to be viewed as a success. Refiners, initially lukewarm about this alternative, have made good use of this program. It stands out amidst a stream of incentive-based programs as the "noble" exception in that it conforms most closely to the economist's notion of a smoothly functioning market.

Given the success of this market in promoting cost savings over a period in which lead was being reduced, it is important to understand why the market was successful. The lead market has two important features that distinguish it from other markets in environmental credits. The first, noted above, was that the amount of lead in gasoline could be easily monitored with the existing regulatory apparatus. The second is that the program was implemented after agreement had been reached about basic environmental goals. In particular, there was already widespread agreement that lead was to be phased out of gasoline. What this suggests is that the success in lead trading may not be easily transferred to other applications in which monitoring is a problem, or environmental goals are poorly defined. Nonetheless, the fact that this market worked well provides ammunition for proponents of market based incentives for environmental regulation.

3.2.2.4. New Directions for Marketable Permits

An interesting potential application for marketable permits has arisen in the area of nonpoint source pollution.[25] The state of Colorado has recently implemented a program which would allow limited trading between point and nonpoint sources for controlling phosphorous loadings in Dillon Reservoir (Elmore et al., 1984). The state and local water agencies are in charge of administering the program. The primary purpose of the program is to improve water quality in the reservoir. The program was implemented in 1984. Firms receive an allocation based on their past production and the holding capacity of the lake. At this point in time, no trading

---

[25] Point sources represent sources that are well-defined, such as a factory smoke stack. Non-point sources refer to sources whose emission points are not readily identified. Examples include fertilizer run-off from farms, and water pollution resulting from contamination by animals. Non-point sources are typically more difficult to control than point sources.

between point and nonpoint sources has occurred. However, trading is expected to take place over the next five years as treatment plants and developers need to obtain credits for phosphorous (Overeynder, 1987).

As in the case of emissions trading, point sources are required to make use of the latest technology before they are allowed to trade. The conventional permitting system is used as a basis for trading. Moreover trades between point and nonpoint sources are required to take place on a 2 for 1 basis. This means for each gram of phosphorous emitted from a point source under a trade, two grams must be reduced from a nonpoint source (Elmore *et al.*, 1984, p. 14).[26] Annual cost savings are projected to be about $800,000 (Kashmanian *et al.*, 1986, p. 14); however, projected savings are not always a good indicator of actual savings, as was illustrated in the case of the Fox River.

The applications covered in this section illustrate that there are a rich array of mechanisms that come under the heading of emission charges and marketable permits. Table III provides a summary of these applications. The table is divided into two parts. The first part reviews the experience with specific charges; the second part summarizes various applications of marketable permits. The table illustrates that there is a great deal in common within each instrument class, including the purpose of implementing the instrument, the effect on cost savings, the effect on environmental quality, and the basis for allocating property rights.

One of the striking features of these instruments is that only a handful of the existing options provide incentives for firms to reach an ambient standard in a manner which significantly reduces overall costs. The charge systems are generally designed to raise revenues for specific activities. The marketable permit approaches, while encouraging more efficient forms of pollution control, have for the most part failed to live up to their theoretical potential. The performance of these activities can yield important clues about the potential for judicious development and application of these mechanisms to new problem areas.

---

[26] See Hahn (1987c) for a discussion of the implications of using trading rules that discriminate among different types of sources.

TABLE III
Examples of Charges and Marketable Permits

Part 1: Charges

	France	Germany	Netherlands	United States
Location:	France	Germany	Netherlands	United States
Instrument:	Effluent and User Charges	Effluent Charges	Effluent Charges	User charges levied by waste-water utilities
Medium:	Water	Water	Water	Water
Targets:	TSS, BOD, and COD, soluble salts, toxics, sewage	Settleable solids, COD, BOD, Cadmium mercury and toxicity for fish	COD, nitrogen, heavy metals such as cadmium, and suspended solids	BOD: COD: TSS, Volume
Measure:	Expected volume and concentration; sometimes actual	Expected volume and concentration	Actual amounts, except for small firms and households	Volume and strength charges; based on individual and average behavior
Target Groups:	Industry and consumers	Industry, municipalities	Industry and consumers	Industry and consumers
Primary Jurisdiction:	Basin Agency	Lander (states)	Water boards; Ministry of Transport and Public Works	Local areas with plants
Principal Purpose(s):	Raise revenues; subsidize treatment and sewerage projects	Raise revenues	Raise revenues	Help meet revenue requirements
Date of Implementation:	1969	January, 1981 (nation-wide)	1969	EPA regulations encourage from 1972
Activity Level:	Widespread	Widespread	Widespread	Widespread application
Cost Savings:	NA	NA	NA	Apparently small
Environmental Quality:	NA	NA	Marked improvement	Little change

Part 2: Marketable Permits

	United States	United States	Fox River, Wisconsin	Dillon Reservoir, Colorado
Location:	United States	United States	Fox River, Wisconsin	Dillon Reservoir, Colorado
Instrument:	Emissions Trading	Lead Trading	Marketable Discharge Permits	Marketable Permits
Medium:	Air	Air	Water	Water

		Lead	BOD	Phosphorous
Environmental Targets:	Volatile organic compounds, carbon monoxide, sulfur dioxide, particulates, nitrogen oxides			
Measure:	Quantities of emissions, based on actual behavior	Quantity of lead in gasoline	Permits defining allowable discharges during different times of the year	Phosphorous loadings
Target Groups:	Industry	Refiners	Pulp and paper industry, municipal treatment plants	Point and nonpoint sources of phosphorous; industry, developers and treatment plants
Primary Jurisdiction:	Varies by Program—primarily state and local with guidelines set by EPA	Federal	Wisconsin	State and local water agencies subject to federal guidelines
Principal Purpose(s):	Promote flexibility in meeting air quality objectives, reduce costs, speed attainment	Allow firms greater flexibility while meeting more stringent standards	Allow firms greater flexibility while preserving environmental quality	Improve water quality in reservoir
Date of Implementation:	1974—netting; 1976—bubbles; 1979—offsets	Inter-refinery trading, 1982; Banking of lead credits, 1985	1981	1984
Amount of Activity:	Moderate—See Table II	High	1 trade	None
Cost Savings:	Substantial—See Table II	Difficult to estimate, but market is very active	Small	None
Environmental Quality:	Little change	No change	No change	No change
Allocation Method:	Grandfathering	Based on current standard and production levels	Grandfathering	Grandfathering
Trading Restrictions:	Technology-based standards, new source requirements, trading ratio	Minimal, must trade within quarter or bank	Must justify need, new source requirements	Technology-based standards, 2:1 trading ratio

NA—not available; BOD – biological oxygen demand; COD—chemical oxygen demand; TSS—total suspended solids.

SOURCES: Boland (1986), Bower et al. (1981), Bressers (1983), Brown (1984a), Brown and Bressers (1986), Brown and Johnson (1984), Elmore et al. (1984), Hahn and Hester (1986), Liroff (1986), Novotny (1986), Patterson (1987), U.S. EPA (1985a).

3.3. Lessons to be Learned

Applications of the textbook definition of charges and marketable permits are all but nonexistent in the real world. It is instructive to explore how actual applications depart from the ideal. This endeavor will help provide the basis for a more informed analysis of policy alternatives. The first part of this analysis develops generalizable insights related to the implementation of charges and marketable permits. The second part examines how the theory of instrument choice can benefit by incorporating some of these insights. The third part assesses the implications of this analysis for system design and performance, and offers some predictions about the future of incentive-based approaches.

3.3.1. General Patterns in Design and Performance

One of the key themes in the application of charges relates to their underlying purpose. *The major motivation for implementing emission fees is to raise revenues.* These revenues are almost always earmarked for activities that promote environmental quality. Usually, the revenues are used to provide subsidies, grants or loans to private firms and to help support treatment plants. While the cases presented in the table are not designed to be representative, almost all environmental charges that have been studied in detail appear to be based on a desire to raise revenues. This is not to say that incentive effects are not important in some cases. However, they are not usually the driving force behind charges. Moreover, up to this point, the number of charges that have documented effects on polluter behavior is only a small subset of the total number of charges.[27] There are several reasons why existing charges have not had a marked effect on firm behavior. The first, and probably most important, is that most charges are not large enough to have a dramatic impact on the behavior of polluters. A second reason is that charges are not always designed to have such an effect. In particular, many charges are not directly related to the behavior of individual firms and consumers. As noted in the table, many are based on expected behavior which often is based on the behavior of

---

[27] Perhaps the best known case is that of the Netherlands (Brown and Bressers, 1986). However, there are others. For example Barde (1986, p. 9) notes that Sweden imposed a tax on fertilizers that resulted in a drop in consumption.

the "average" firm in an industry. A third reason relates to the unavailability of data necessary to test the hypothesis that charges have led to environmental improvements.

While charges have not had a major effect on incentives, they are widely perceived to have had a positive impact on environmental quality. To the extent that the revenues from charges are ear-marked for pollution control equipment or refinements in process technology, this is not totally unexpected. Indeed, *the application of charge revenues to abatement activities is the primary mechanism through which charges address environmental quality.* Direct incentive effects are much less pronounced.

Another feature of charges which is important in terms of their implementation is that *there is a tendency for the charge to increase over time,* even accounting for inflation. Presumably, starting out with a relatively low charge is a way of managing political opponents, and determining whether the instrument will have the desired effects.[28]

The application of marketable permits presents an interesting contrast to that of charges. The *primary motivation behind marketable permits is to provide increased flexibility in meeting prescribed environmental objectives.* This flexibility, in turn, allows firms to take advantage of opportunities to reduce their expenditures on pollution control without sacrificing environmental quality. In the case of emissions trading and lead trading, marketable permits appear to have given rise to substantial cost savings. In the case of the Fox River and Dillon Reservoir, trading has been limited. The performance of these markets is very much tied to the rules that govern their operation. In the case of emissions trading and permit trading on the Fox River, the level of regulatory involvement in individual trades is quite high. Moreover, there are many restrictions on trading. Both of these features tend to limit the scope for trading, and one of the features observed in both these markets is that their performance has fallen far short of their theoretical potential. In contrast, the lead trading market has enjoyed vigorous levels of activity, and appears to have been very

---

[28] A good analogy to the example of charges is the introduction of the U.S. federal income tax, which started at very nominal levels and then was raised over time. Note, however, that the increases don't necessarily go on forever (thankfully).

successful, though hard data on the relationship between actual and cost savings are not readily available.

An interesting feature of the major marketable permit programs is that they go to some lengths to deemphasize the explicit nature of the property right. In the case of emissions trading, rights are defined in terms of "credits." In the case of lead trading, rights are defined in terms of "inter-refinery averaging." The subtlety in the definition of these activities reflects the fact that environmentalists are likely to be quite sensitive to how these trading programs are advertised. This suggests that the packaging and marketing of market-based reforms can have an important effect on whether they are adopted.

A careful examination of the charge and marketable permits schemes reveals that they are rarely, if ever, introduced in their "pure" form. *Virtually all environmental regulatory systems using charges and marketable permits rely on the existing permitting system.* This result should not be terribly surprising. Most of these approaches were not implemented from scratch. Rather, they were grafted onto regulatory systems in which permits and standards play a dominant role. While it is true that these alternative approaches are implemented in conjunction with other instruments, it is also important to recognize that the charge and marketable permits systems differ in the extent to which they rely on the existing regulatory apparatus. However, all the systems examined here rely very heavily on the existing regulatory system.

Moreover, the systems are linked to the existing regulatory approach in another important respect. *Virtually all of the incentive-based systems which are used focus on individual pollutants.* In many cases, this approach is justified. However, there are many pollution problems for which synergisms can be quite important. For example, the formation of sulfates which contribute to acid deposition is integrally related to emissions of nitrogen oxides and volatile organic compounds (Hahn, McRae and Milford, 1987). Another more well-known case is the formation of ozone which results from emissions of nitrogen oxides and volatile organic compounds. Interestingly, some areas using emissions trading are beginning to experiment on a limited basis with interpollutant trades. However, this is a highly contentious issue.

At this point in time, there is a marked difference in the relative utilization of the two types of instruments. *Charges currently enjoy more widespread use than marketable permits.* Admittedly, it is hard to measure utilization. However, charges are currently used in more individual applications, in more countries, and for more pollutants than marketable permits. It is hard to know whether this trend will continue as experience is gained with both of these instruments.[29]

The level of cost savings resulting from implementing charges and marketable permits is generally far below their theoretical potential. Cost savings can be defined in terms of the savings which would result in meeting a prescribed environmental objective in a less costly manner. As noted earlier, most of the charges to date have not had a major incentive effect. We can infer from this that polluters have not been induced to search for a lower cost mix of meeting environmental objectives as a result of the implementation of charge schemes. Assuming the current regulatory system departs a great deal from the "ideal" solution yields the result that charges have not performed very well on narrow efficiency grounds.[30] The experience on marketable permits is more direct. Hahn and Hester (1986) argue that cost savings for emissions trading fall far short of their theoretical potential. The only apparent exception to this observation is the lead trading program, which has enjoyed very high levels of trading activity.

The example of lead trading leads to another observation about the efficiency of different charge and marketable permit systems. In general, *the two instruments exhibit wide variation in their effect on economic efficiency.* Some charges have a marked effect on the generation and disposal of pollutants; others do not. Similarly, some marketable permit approaches have led to significant cost savings while others have not. On the whole, there is more evidence for cost savings with marketable permits than with charges.

---

[29] There appear to be some important differences in the application of charges across media, with charges being used more widely for managing water than air. While Table I does not provide overwhelming support for this view, recall that Table I does not measure the intensity of the various applications. The major revenues from charges appear to result from water-related cleanup activities. For example, in the Netherlands, revenues from water-related charges account for approximately 85% of total revenues from charges (Opschoor, 1986, p. 3).

[30] Evidence supporting the view that the current U.S. system departs from the ideal is presented in Tietenberg (1985). While Tietenberg's discussion pertains to the U.S., there is little reason to believe that the systems adopted in other countries are very different in terms of the emphasis placed on identifying economically efficient solutions to environmental problems.

The charge systems and marketable permit systems that have been implemented have behaved in a manner that is consistent with economic theory. This observation may appear to contradict what was said earlier about the departure of these systems from the economic ideal. However, it is really an altogether different observation. It suggests that the performance of the markets and charge systems can be understood in terms of basic economic theory. For example, where barriers to trading are low, more trading is likely to occur. Where charges are high and more directly related to individual actions, they are more likely to affect the behavior of firms or consumers. At this point, our knowledge of the behavior of environmental instruments is largely anecdotal. One challenge for future research will be to gain a better understanding of the economic response to different instruments.

Another important measure of the performance of these instruments is their effect on environmental quality. In general, the effect of both charges and marketable permits on environmental quality appears to be neutral or positive. The effect of lead trading has been neutral in the aggregate. The effect of emissions trading on environmental quality has probably been neutral or slightly positive. Charges can affect environmental quality both directly, through inducing firms to cut back on pollution, and indirectly, by being used to subsidize abatement activities. The direct effect of charges has been modest. Charges are generally designed to promote environmental quality through the redistribution of funds to these activities. The indirect effect of charges on environmental quality has been significant.

The evidence on charges and marketable permits points to an intriguing conclusion about the nature of these instruments. Charges and marketable permits have played fundamentally *different* roles in meeting environmental objectives. *Charges are used primarily to improve environmental quality by redistributing revenues. Marketable permits are used primarily to promote cost savings.*

3.3.2. Towards a More Complete Theory of Instrument Choice
The formal theory on the choice of instruments is noteworthy for its simplicity. It starts from the premise that bureaucrats and legislators choose specific instruments to further their own objectives. The

basic insight of these theories is that the choice of instruments will be affected in systematic ways by individual perceptions about the winners and losers from various policies (Campos, 1987; McCubbins and Page, 1986).

The theory of instrument choice, as it relates to pollution control, has been greatly influenced by the work of Buchanan and Tullock (1975). These authors attempted to explain the preference for standards over emissions charges, and did so using a simple argument related to the rents an industry would receive from cartelization. While this argument is elegant, it misses two important points. The first is that within particular classes of instruments, there is a great deal of variation in the performance of instruments. The second is that most solutions to problems involve the application of multiple instruments. Thus, while the theory explains why standards are chosen over an idealized form of taxes, it does little to help explain the rich array of instruments that are observed in the real world. In particular, under what situations would we be likely to observe different mixes of instruments?

Another weakness in the existing theory is that the instruments fail to behave in a way that is suggested by the theory. The preceding analysis points to some striking similarities and differences in the use of these mechanisms. Most emissions charges are used as a revenue raising device for subsidizing abatement activity. Yet, a few also have pronounced direct effects on polluters. Most marketable permit approaches are designed to promote cost savings while maintaining environmental quality. However, existing approaches reveal very different performance characteristics in terms of the degree to which they yield cost savings.

While a more formal reconstruction of these theories will be taken up in the next section, it is instructive to examine how the patterns in the application of these incentive-based approaches can be used to construct a more coherent theory of instrument choice. One of the key elements in the positive theory is the focus on distributional considerations. This focus is well-placed in that distributional concerns can often provide important clues about the likely range of feasible choices. In the case of charges, it is clear that distributional concerns play an important role in acceptability. While the use of revenue is rarely linked to individual contributions, it is usually earmarked for environmental activities related to

those contributions.[31] Thus, for example, charges from a noise
surcharge will be used to address noise pollution. Charges for water
discharges will be used to construct treatment plants and subsidize
industry in building equipment to abate water pollution. This trend
suggests that different industries want to make sure that their
contributions are used to address pollution problems for which they
are likely to be held accountable. Thus, industry sees it as only fair
that, as a whole, they get some benefit from making these
contributions. Individual firms in an industry need not receive
benefits in direct proportion to their contributions, but overall
contributions will generally be earmarked for that industry.

The "recycling" of revenues from charges points up the importance of the existing distribution of property rights. This is also true
in the case of marketable permits. The "grandfathering" of rights to
existing firms based on the current distribution of rights is an
important focal point in many applications of limited markets in
rights (Rolph, 1983; Welch, 1983). The examples considered here
are consistent with this observation. Emissions trading and the
trading of permits on the Fox River place great importance on the
existing distribution of rights. Lead trading places somewhat less
importance on the existing distribution of rights in that it does not
directly grandfather rights to existing participants. Nonetheless, it
does use current standards as a baseline for determining how rights
are allocated.

The bottom line is that *all of these systems place great importance
on the status quo.* Charges, when introduced, tend to be phased in.
Marketable permits, when introduced, usually are optional in the
sense that existing firms can meet standards through trading of
permits or by conventional means. In contrast, new or expanding
firms are not always afforded the same options. For example new
firms which use state-of-the-art equipment must still purchase
emission credits if they choose to locate in a nonattainment area.
This is an example of a "bias" against new sources. It results from
the fact that new sources have less of a claim on current wealth than
existing sources. Consequently, they are asked to pay a higher price
in terms of the environmental costs they will face. While not

---

[31] In his study of the Netherlands, Opschoor (1986, p. 22) notes that all charge
revenues are earmarked.

efficient from an economic viewpoint, it is consistent with the political reality that new sources don't "vote" and existing sources do.

Though the *status quo* is important in all applications studied here, it does not, by itself explain the rich variety of instruments that are observed. The fact that instruments perform differently in terms of both their performance and distribution needs to be explained. To date, very little work has been done in this area.

The *status quo* defines a distribution of wealth among participants in a political process. It does not speak directly to the underlying preferences of groups and individuals who have a stake in the outcome. By examining these underlying preferences, it may be possible to gain further insights into the nature of instrument choice. Hahn and Hester (1986) have argued that the case of emissions trading can best be understood in terms of a struggle over the underlying property rights. This insight can be extended to the comparison of the lead program and the emissions trading program. The performance of these two programs is directly related to the degree of controversy over the underlying distribution of property rights.

There has been heated controversy over emissions trading since its inception. Several environmental groups have made continued attempts to thwart the expansion of this option. In contrast, there has been comparatively little controversy over the implementation of lead trading. How can be begin to understand the difference in attitudes towards these two programs, both of which had their origins at the federal level in the U.S.?

There are several important differences between these two programs. In the case of lead standards, there appears to be agreement about the distribution of property rights, and the standard that defined them. Refiners had the right to put lead in gasoline at specified levels during specified time periods. Lead in gasoline was reduced to a very low level by the end of 1987. In contrast to lead, there is great disagreement about the underlying distribution of property rights regarding emissions trading. Environmentalists continue to adhere to the symbolic goal of zero pollution. Industry believes and acts as if its current claims on the environment represent a property right.

In the case of lead trading, output could be relatively easily monitored using the existing regulatory apparatus. This was not so for the case of emissions trading. A new system was set up for evaluating proposed trades. This was, in part, due to existing weaknesses in the current system of monitoring and enforcement. It was also a result of concerns that environmentalists had expressed about the validity of such trades.

The effect that emissions trading was likely to have on environmental quality was much less certain than that of the lead trading program. Some environmentalists viewed emissions trading as a loophole by which industry could forestall compliance. Indeed, there is evidence that some firms may have used bubbles to avoid compliance deadlines (Hahn and Hester, 1986). The effects of lead trading were much more predictable. Until 1985, there was no banking, so the overall temporal pattern of lead emissions would remain unchanged under the program. With the addition of banking in 1985, this pattern was changed slightly, but within well-defined limits.

To accommodate these differing concerns, different rules were developed for the two cases. In the case of lead trading, rights are traded on a one-for-one basis. In contrast, under emissions trading, rights are not generally traded on a one-for-one basis. Rather, each trade must show a net improvement in environmental quality. In the case of lead, all firms are treated equally from the standpoint of trading. In the case of emissions trading, new firms must meet stringent standards before being allowed to engage in trading.

What this comparison suggests is that it is possible to gain important insights into the likely performance and choice of instruments by understanding the forces that led to their creation. Moreover, the general analysis of instruments presented in this chapter has some important implications for existing theories of instrument choice. First, it shows that the view of the choice as dichotomous, (e.g., as between standards and fees), is unnecessarily simplistic. It may very well be the case that the choice can be viewed as continuous for the purpose of theory. For example, in the cases of lead trading and emissions trading, one of the choice variables might be the extent to which market forces should be used for allocating rights. Moreover, the analysis suggests that efficiency (measured in purely economic terms) may be the by-product of the degree to which groups agree on the underlying distribution of property rights.

This view of efficiency is similar to, but should not be confused with, the notion of efficiency advanced by Becker (1983). Becker argues that government will tend to choose mechanisms that are efficient in redistributing revenues from less powerful to more powerful groups. To the extent that his argument is testable, I believe it is not consistent with the facts. For example, the U.S. currently has a policy that directs toxic waste dumps to be cleaned up in priority order. The policy makes no attempt to examine whether a greater risk reduction could be attained with a different allocation of expenditures. Given a finite budget constraint, this policy does not make sense from a purely economic viewpoint. However, it might make sense if environmentalists hoped that more stringent policies would emerge in the future. Or it might make sense if Congress wants to be perceived as doing the job "right," even if only a small part of the job gets done.

A second example can be drawn from emissions trading. It is possible to design marketable permit systems that are more efficient and that ensure better environmental quality over time (Hahn and Noll, 1982a). Yet, these systems have not been implemented for a variety of reasons. Environmentalists may be reluctant to embrace market alternatives because they fear it may give a certain legitimacy to the act of polluting. Moreover, they may not believe in the expected results. Thus, for Becker's theory to hold in an absolute sense, it would be necessary to construct fairly complicated utility functions. The theory is perhaps better viewed as a tendency of systems. Even as a tendency, however, it may be questioned since it does not explicitly address how choices are made by legislators and bureaucrats.

Some of the literature on instrument choice begins with the objective of legislators. For example, Campos (1987) models the instrument choice problem in terms of maximizing a support function for a single legislator. While this approach is elegant, it ignores the role of bureaucracies in selecting instruments.[32] In the case of environmental applications, specifically charges and marketable permits, many of these ideas originated in the bureaucracy,

---

[32] One advantage of the approach taken by Campos is that it provides a nice way of comparing the relative efficiency of policies that are selected by legislators. Up to this point, most of the political science models merely suggest that policies selected by governments will not be efficient. Identifying the degree of inefficiency is difficult both in theory and practice.

and some were implemented without formal authorizing legislation. Whether this was in line with legislative preferences is not easy to infer. The point, however, is that it is foolhardy not to consider bureaucracies as having an important role in the development and design of instruments.

In short, existing theories could benefit from more realism. This realism could be introduced by a more careful examination of actual applications of instruments. This means that the mixed nature of policies would need to be explained. In addition, the instrument choice problem needs to be defined more carefully. In the case of the environment, marketable permits and charges are only a small part of environmental regulation. Subsidies and standards represent a major part of this system, and thus would need to be included in a more comprehensive theory of policy choice.

The political economy of instrument choice can benefit from the fact that these choices have now been studied in a variety of contexts. As can be seen from Table I, a large array of countries use fees; however, only two countries use marketable permits. Moreover, the application in Germany is fairly limited. How can this differential usage of instruments across countries be explained? Noll (1983) has argued that the political institutions of different countries can provide important clues about regulatory strategy. In addition to institutional structure, which is undoubtedly important, there are other issues that can play a key role. The comparison of lead trading and emissions trading revealed that the very *nature* of the environmental problem can have an important effect on interest group attitudes.

Interest group attitudes can be expected to vary across countries. In the Netherlands, Opschoor (1986, p. 15) notes that environmental groups tend to prefer charges while employer groups prefer regulatory instruments. Barde (1986, pp. 10–11) notes that the political "acceptability" of charges is high in both France and the Netherlands. Nonetheless, some French airlines have refused to pay noise charges because the funds are not being used (Barde, 1986, p. 12). In Italy, there has been widespread opposition from industry and interest groups (Panella, 1986, pp. 6, 22). While German industry has accepted the notion of charges, some industries have criticized the differential charge rates across jurisdictions. In the United States, environmentalists have shown a marked preference for regulatory instruments, eschewing both charges and marketable permits. These preferences may help to explain the choice of instruments in various countries as well as the relative utilization of different instruments.

The choice set will also be affected by what is known about the performance of different instruments. Thus, channels for exchanging information can play an important role on both the timing of regulation and the strategy for regulation. For example, recent applications of charges in OECD countries may have resulted in part from the success of earlier applications in other OECD countries. The role of information and ideas in determining the feasible space of alternatives is only just beginning to be appreciated in the area of regulation (e.g., see Derthick and Quirk, 1985; and Brickman, Jasanoff and Ilgen, 1985). It is quite clear that convincing real world applications or "experiments" can have important impacts on the design of new policies. For example, in the case of airlines, there was persuasive evidence from earlier experience with intrastate competition that helped pave the way for U.S. deregulation (Bailey, 1986; Levine, 1981).

3.3.3. Implications for System Design and Performance

The review of marketable permits and charge systems has demonstrated that regulatory systems involving multiple instruments are the rule rather than the exception. The fundamental normative problem is to determine the most appropriate mix. This mix should be selected in light of the political realities that have shaped the implementation of various tools.

Brown and Johnson (1984, p. 946) have argued that, in general, more policy instruments are preferred to less. This is clearly true in a situation where policy instruments are costless to implement. In the real world, however, instrument choice requires tradeoffs among several objectives (Hahn, 1986). Moreover, the argument by Brown and Johnson implicitly assumes that policies will be used to achieve a given objective. When objectives of regulatory strategy come into conflict, the situation is much less clear (Lave, 1984). Even when regulatory strategies do not come into conflict, more instruments may not be preferred to less.[33] For example, the Netherlands has decided to combine several charges into one fuel

---

[33] Barde (1986, p. 23) argues that too great a variety of tools may limit their usefulness.

charge as a means of simplifying the charge system (Opschoor, 1986, p. 22). At this point, all that can be safely concluded is that there is no simple relationship between the number of instruments used and the effectiveness of a policy.

In addition to selecting an appropriate mix of instruments, attention needs to be given to the effects of having different levels of government implement selected policies. The appropriate level of government will be dictated by several factors. One important factor is the scope of the problem. If the problem is local, then the logical choice for addressing the problem is the local regulatory body. However, this is not always true. For example, the problem may require a level of technical expertise that does not reside at the local level, in which case some higher level of government involvement may be required. What is clear from a review of implementing environmental policies is that the level of oversight can affect the implementation of policies. For example, Hahn and Hester (1986) note that a marked increase in bubble activity is associated with a decrease in federal oversight.

The problems of choosing an appropriate level of government for addressing problems and choosing an appropriate menu of instruments for resolving issues are not new. What is new is the information we can now bring to bear on these issue through systematic empirical inquiry into the effects of various instruments and implementation strategies.

The evaluation of existing incentive-based mechanisms has some important implications for their use in the future. Because marketable permit approaches have been shown to have a demonstrable effect on cost savings without sacrificing environmental quality, this instrument can be expected to receive more widespread use.[34] One factor that will stimulate the application of this mechanism is the higher marginal costs of abatement that will be faced as environmental standards are tightened. A second factor that will tend to stimulate the use of both charges and marketable permits is a "demonstration effect." Several countries have already implemented these mechanisms with encouraging results. The experience gained in implementing these tools will stimulate their use in future

---

[34] The recent EPA regulations controlling chlorofluorocarbons and halons that contribute to the depletion of stratospheric ozone provide a case in point (Hahn and McGartland, 1988).

applications. A third factor that will affect the use of both of these approaches is the technology of monitoring and enforcement. As monitoring costs go down, the use of mechanisms such as direct charges and marketable permits can be expected to increase.

In general, the prediction is that greater use of these systems will be made in the future. The "demonstration effect" will be critical in decreasing opposition from industry and environmental groups. Initially, industry has opposed many structural changes in environmental regulation, including both charges and marketable permits. This resistance can be explained, in part, by a type of risk aversion. People and businesses already successfully operating in one environment are often reluctant to make changes that could dramatically affect the "rules of the game" (Allison, 1971; Simon, 1976).[35] Nonetheless, as experience with these approaches grow, they are likely to receive greater support, unless they can be shown to have demonstrable adverse consequences.

3.4. Conclusions and Areas for Future Research

The preceding analysis reveals that economic approaches to environmental reform are much more complicated than had initially been theorized. However, the performance of both sets of instruments examined here is broadly consistent with economic theory. In the case of permit trading, the performance of the market was integrally related to the rules governing the market. In the case of charges, it appears that incentive effects are important in a few applications. In general, charges have tended to serve a revenue-raising function. Marketable permits have served primarily as a mechanism for promoting cost savings.

The study of the politics surrounding environmental reform is in its infancy. This section argued that the existing distribution of property rights has a major impact on system design. It also argued that the existing regulatory structure has a major impact on design. Most of the systems examined here can be viewed as *incremental* departures from the *status quo,* in that they make liberal use of the existing regulatory apparatus.

---

[35] This resistance to change is not restricted to environmental regulation. See, for example, the discussion of airline deregulation in Derthick and Quirk (1985).

There is a great deal of work that will be required to construct a more general theory of instrument choice—even a theory that is limited to environmental or social regulation. A useful starting point is to begin to gather data on the political and economic effects of different environmental policies. This section has examined charges and marketable permits. These two instruments represent only a small subset of the actual instruments that are used to motivate firms to clean up the environment. And even for these two cases, our knowledge is quite limited. Subsidies, loans and grants are also very important; yet, the quantitative impacts of these policies are not well understood. Moreover, little attention has been given to sorting out the effects that different tools will have when taken together as a whole.

Because of the increase in new regulatory approaches for managing environmental problems, the time is ripe to broaden the political focus of existing research. One challenge is to identify and explain common themes that exist in different countries. An example that frequently appears in U.S. regulation is the asymmetric treatment given to new and existing sources. It would be useful to know the extent to which new sources are generally asked to meet more stringent requirements than existing sources. This might be a policy that is *invariant* across political systems, and it would be useful to know, both for purposes of political theorizing and designing new policies, whether this is indeed the case.

The issue of political feasibility can also be fruitfully addressed through a more careful assessment of the positions of key interest groups in the political process. Very little of the work on charges attempts to link these approaches to underlying political interests. Understanding why charges are selected, and why they differ dramatically across jurisdictions and countries is an important challenge for future research.

Another important area for research is to examine the effects of information transfer on policy choice. Few authors have addressed its importance, but it is clear that the process by which information is disseminated is likely to have important effects on the timing of regulations, what gets regulated, and how problems are regulated.

Lest the agenda for research sound too overwhelming, it is important to take stock of the progress that has been made to date. Beginning from some stylized definitions of policy instruments that were used for decades in the normative literature on environmental economics, and adopted more recently in the positive literature on instrument choice, we are now at the point where we can provide a better characterization of how different instruments perform in practice. We can also provide partial explanations for the performance of existing instruments as well as partial explanations for their selection. The challenge that lies ahead is to provide a more systematic linkage between the theory and the actual performance of instruments.

4. THE POLITICAL ECONOMY OF ENVIRONMENTAL REGULATION: TOWARDS A UNIFYING FRAMEWORK

4.1. Introduction

During the last two decades, there has been tremendous growth in the scope of environmental regulations. More chemicals are regulated than ever before. The stringency of regulations has also increased over time, particularly in densely populated areas. Despite the increase in the level of environmental regulation, the dominant approach to regulation has changed very little. In most places, a central regulatory authority sets standards. These standards vary in type, but they typically place stringent emission limits on individual sources.

In addition to standards, governments have made liberal use of subsidies to help promote environmental quality. For example, the U.S. federal government provided large subsidies, in the billions of dollars, to aid in the construction of municipal waste treatment plants. States often provide subsidies and tax incentives to aid in the control of pollution. Indeed, both standards and subsidies have enjoyed widespread use in most developed countries.

Less widespread, but growing in popularity, is the application of tools that economists find more appealing from an economic efficiency perspective. Examples include effluent fees and marketable permits. While the implementation of these instruments tends to depart substantially from the textbook versions, the application of these tools has had a marked impact on environmental quality and the costs of achieving environmental goals.

Given the vast array of approaches to environmental regulation, it is only natural to ask how their selection might be explained or rationalized. The first step in searching for a deeper understanding is to identify patterns in environmental regulation. The second step is to examine underlying forces that might help explain these patterns. The purpose of this section is to provide simple rationales for many of the patterns that are observed in environmental policy. An understanding of the basic forms that environmental regulation takes will help provide insights into the conditions and potential for regulatory reform.

Several scholars have attempted to grapple with understanding different aspects of environmental policy using positive political theories. This research will review and build on the insights that have been developed. There are two primary contributions of this section. The first is to provide more complete explanations for a number of patterns in environmental policy that have not been satisfactorily explained. For example, there is, as yet, no satisfactory theory about the emergence of incentive-based mechanisms, such as marketable permits and effluent fees. This section develops some formal models that shed light on these issues. The second contribution of this section is to develop a parsimonious framework for understanding many important aspects of environmental policy. This framework views the outputs of environmental policy as emerging from a struggle between key interest groups.

Positive theories pertaining to the application of environmental regulation are critically examined in part 4.2. Part 4.3 presents a formal analysis aimed at identifying key factors that affect policy design. A series of models are presented that provide insights into the existing standard-setting process, new regulatory approaches, and dominant patterns in environmental policy. Part 4.4 raises some broader issues related to the construction of a theory of instrument choice. Finally, part 4.5 reviews the key conclusions and suggests areas for future research.

4.2. Theories and Explanations: A Critical Appraisal

Before discussing existing positive theories of environmental regulation, it is useful to identify suitable goals for a positive theory. These goals, taken in conjunction with the existing state of the art, will help to suggest a research agenda.

Ideally, what would we want from a positive theory of

environmental regulation? Like any positive theory, we would hope that it has predictive power. Moreover, it should be able to explain what gets regulated, the methods chosen for regulation, and the likely winners and losers from regulation. While there are other items that a philosopher-king or a scientist might wish to know, this list is already forbidding enough. Indeed, at present, there is very little in the way of theory that suggests what will be regulated. No doubt, the choice of what is regulated is greatly affected by the state of scientific knowledge. However, the state of scientific knowledge does not suggest precisely which carcinogens will be regulated; nor does it suggest whether a large number or a small number of carcinogens will be regulated. While some insights from political theory can be brought to bear on these questions, our knowledge of what will get regulated is quite limited.[36]

In contrast to our rudimentary understanding of what gets regulated, our understanding of how things get regulated, and the associated distribution of benefits and costs, is relatively advanced. Indeed, virtually all of the positive theories that have been developed are based on some notion of net benefit maximization either by a single actor or in the context of a mathematical game. Most of these theories were reviewed in Section 3. The basic insight is that the choice of a regulatory instrument depends crucially on the specific instruments that are compared. It also depends on the amount of power that particular interest groups have, and how this power is wielded in the political process.

The influence of different interest groups has been modeled in several ways. Perhaps the most popular (and also the most tractable) is to assume that a single agent, such as a legislator, chooses policies to maximize net benefits. The initial framework for this maximization problem was suggested by Peltzman (1976) in the context of regulation. Campos (1987) extended this framework to the instrument choice problem (see Section 3). In another applica-

---

[36] Several authors have suggested that information transfer plays an important role in determining what gets regulated in various countries (e.g., see Badaracco, 1985, and Dowlatabadi and Hahn, 1986). However, the role of these mechanisms is poorly understood. Moreover, it tends to beg the question of why things get regulated in the first place. Many people often point to "crises" as motivating regulation. While crises certainly induce responses on the part of government officials, it would be nice if we had a theory of crisis formation along with an understanding of how different crises lead to different responses.

tion of Peltzman's approach, Magat, Krupnick and Harrington (1986) attempted to explain how different groups affect the stringency of standards at different points in the rulemaking process.

While some theories of instrument choice are based on direct redistribution from one group to another, there are others that build on the political decisions to delegate power, and the form of the power that is delegated. Both Fiorina (1982) and McCubbins (1985) have focused on Congress, and attempted to identify conditions under which policies will be delegated. These models are important because they accentuate the role of the legislature in determining the nature of policy. McCubbins and Page (1986) illustrate how many of these ideas can be applied to environmental policy. The authors argue that economic incentive schemes may not be selected because they tend to increase conflict and uncertainty among politicians by providing firms with greater flexibility. While this may be true, there are many instances in which the government has opted to use economic incentive schemes for both social and economic policies. Moreover, the use of these approaches is becoming more widespread. Thus, some further explanation about the emergence of these schemes would be helpful.

There have been several stories and theories about the winners and losers from environmental policy. Tucker (1982) argues that the environmental movement in the U.S. primarily serves to enhance the wealth of the privileged class. Ackerman and Hassler (1981) paint a somewhat different picture. Studying the emergence of regulations that required power plants to install scrubbers, the authors argue that a coalition formed among environmentalists and eastern coal interests. The resulting regulations were very expensive and may actually have resulted in lower environmental quality. The case study is important because it provides evidence that environmental groups may be more concerned with symbols, such as forced scrubbing, than actual environmental outcomes. It also shows how interest groups can form coalitions that yield seemingly bizarre outcomes, yet are perfectly sensible from the viewpoint of the interest groups involved.

There is a debate in the published literature about the extent to which new regulations benefit well-organized interests. Stigler (1971) argues that producers will generally be the beneficiaries of regulation. Rolph (1983) takes issue with this finding, arguing that the existing distribution of property rights is important in shaping new policies, but not finding a systematic trend for new regulations to favor well-organized interest groups such as producers. Certainly in the area of environmental regulation, the verdict on this issue is out. Welch (1983) argues that the current distribution of property rights strongly affects the design of new incentive-based policies. The examination of incentive-based programs in Section 3 provides strong support for the view that the existing distribution of property rights has a significant effect on program design. However, the major beneficiaries of these policies seem to vary from case to case. This should not be particularly surprising since the configuration of interest group influence can also be expected to vary.

Formal tests of theories about the beneficiaries of environmental regulation are just beginning to emerge. Perhaps, the best known theory involves the use of standards to enhance industry profits. After laying out the theory, Maloney and McCormick (1982) present some empirical support based on cotton dust standards and an air pollution ruling affecting smelters. While they argue that the results are consistent with the view that industry benefits from regulations, a detailed analysis of the cotton dust case by Hughes, Magat, and Williams (1986) casts doubt on their conclusions.

Though this section focuses on formal explanation of instrument choice, it is important to note that there has been no dearth of casual explanations that purport to explain various aspects of the instrument choice problem. Perhaps, the problem that has provided the greatest puzzle for social scientists is why standards are so widely applied in social regulation. One argument not considered here is that standards are more palatable to the professionals who play a major role in making regulatory decisions. Lawyers are accustomed to writing standards that dictate how firms and individuals should behave. Similarly, engineers are used to dealing with specific design criteria for pollution control equipment. A second argument for the use of standards is that they have symbolic value. They often provide the appearance of taking a tough, no-nonsense, attitude towards problem solving. The argument about professional norms seems persuasive, but it is difficult to model. The argument about symbolism is more easily modeled, and will be explored in more detail later in this chapter.

This review of the state of our understanding of environmental policy choice reveals that virtually all the formal models are based on theories about redistribution and power. The simplest models assume that industry has all the power, and that there is a single decision maker. More elaborate models relax these assumptions. The models help to explain some important stylized facts about the choice of standards over other instruments and the likely beneficiaries of environmental regulation. However, they provide few insights on the conditions under which incentive-based instruments will be chosen, the mechanics of the standard-setting process, and the choice of the form of environmental regulation. The subsequent analysis adds to this theoretical foundation by addressing a variety of issues in environmental policy related to the selection of instruments and the choice of environmental targets.

4.3. Towards a More Unified Theory of Environmental Policy

Environmental policy is almost always at the source of a great deal of controversy. At the heart of this controversy lie two fundamentally opposing points of view. One represented by "industry", usually focuses on the impact of environmental policy on profits. A second, represented by "environmentalists", is more concerned with the impact of policy on the environment. The reduction of the diverse range of interest group perspectives on environmental policy into two distinct viewpoints is a gross oversimplification. The simplification is made purely in the interest of developing a parsimonious and elegant theory of environmental policy. The purpose of this section is to explore how this view of the world can enlighten many of the choices and patterns that are observed in environmental policy. The first part of the section will examine the logic of the standard-setting process. This will be followed by a discussion of the emergence of alternative regulatory mechanisms that address environmental problems. More general themes in the choice of what is regulated and the level of regulation are then identified.

One critical simplification that will be used to facilitate the subsequent analysis is the assumption that environmental policy is made by a single decision maker or decision making unit, typically represented by a regulator or a legislator.[37] While this is clearly at odds with reality, it is again made in the interest of simplicity. Moreover, many of the examples presented here could be recast in the form of a mathematical game in which multiple interest groups compete. The unitary actor assumption is consistent with the models developed by Buchanan and Tullock (1975), Fiorina (1982), and Peltzman (1976). Nonetheless, it suffers from the fact that institutions are not explicitly factored into the analysis. Typically, the institution that most scholars are concerned about factoring into the analysis is Congress (see, e.g. Fiorina, 1982; and McCubbins and Page, 1986). However, for many applications in environmental policy, various levels of bureaucracy also play an important role, and one that has not received the attention it deserves. In addition, in the United States, the courts have played a very important role in the making of environmental policy (Melnick, 1983).

4.3.1. Towards a Theory of Standard-Setting

A useful starting point in addressing issues in instrument choice is to examine how the dominant instrument in environmental policy—the standard—is applied. Surprising as it may seem, there is no generally accepted theory of how environmental standards are applied. Suppose that a regulator is charged with imposing standards on individual sources until a given environmental objective is met. The regulator must decide how standards will be applied. One way to think about this problem is that the regulator must balance economic objectives against political concerns. Suppose there are two types of standards that can be imposed, one that imposes low economic costs on individual firms, and another that imposes high economic costs. The decision in this case is relatively straightforward. Standards with the lower economic cost will be applied first.[38]

---

[37] The motivations of regulators are rarely the same as elected officials. Nonetheless, elected officials can exert a great deal of control over regulators through a variety of oversight mechanisms, such as budget allocations and hearing. In the subsequent analysis, the objectives of the two groups are assumed to be identical.

[38] In the interest of simplicity, the distribution of economic costs across firms is ignored. One view of environmental regulation that has achieved some popularity in the economics literature is that firms use this regulation to increase industry profits or raise rival's costs. There is no question that some firms and industries will try to

TABLE IV
How Standards are Selected by a Regulator

	Low Economic Cost	High Economic Cost
Low Political Cost	1	2
High Political Cost	3	4

Suppose, however, that standards also have a political cost attached to them. This cost might result from standards affecting unemployment, plant closure, or environmental quality in the neighborhood of an important politician. Then the regulator needs to rank standards on two dimensions. Table IV provides a two-by-two matrix representation of the various alternatives facing the regulator. His preference over these alternatives is reflected in the number in each box. The number "1" represents the most preferred alternative and the number "4" represents the least preferred alternative. Clearly, the regulator's first choice is to impose standards with low political and economic costs. Conversely, the least preferred alternative is represented by standards that impose both high economic and political costs. The remaining two cells in the matrix are more difficult to evaluate, and highlight the nature of the balancing problem. Here, it is assumed that political costs dominate economic concerns for the regulator, and thus, a standard with low political costs and high economic costs is preferred to one with high economic and low political costs.

FOOTNOTE 38 (continued)
use the regulatory process in a strategic manner (Maloney and McCormick, 1982; Owen and Braeutigam, 1978). For example, recently the automobile companies and oil companies have been engaged in an argument over who should be required to install control equipment related to reducing automobile emissions. In the interest of brevity, I have chosen not to explore these strategic issues in detail. However, they are certainly important, and they do help to explain some differences between old and new source regulation.

TABLE V
Choice of Standards for New and Existing Sources

		New Sources	
		Low Standard	High Standard
Existing Sources	Low Standard	(1,4) *	(2,2)
	High Standard	(3,3)	(4,1)

* For each ordered pair, the first coordinate represents industry preference, and the second coordinate represents environmentalist preference.

Whether this will always be true depends on the precise nature of the regulator's utility function.

This basic paradigm captures the notion that a regulator needs to balance different concerns; however, it does not explicitly introduce the concerns of interest groups. To explore differences in viewpoints among interest groups, it is instructive to consider a concrete example. One persistent theme in environmental regulation is that new sources of pollution get regulated more stringently than existing sources. A simple reason often used to explain this observation is that new sources don't "vote", while existing sources have access to political power. A slightly different, but complementary, way of looking at this problem is offered in Table V. Suppose a regulator has to choose between low and high standards for new and existing sources of pollution. Instead of showing the preferences of the regulator, Table V shows the preferences of two interest groups. Each ordered pair represents the preferences of industry and environmentalists, respectively. Industry is assumed to prefer low standards across the board, because it reduces costs.[39]

_____

[39] To the extent that new source standards serve as a barrier to entry, industry might value this option more highly. However, this switching of industry preferences does not change the basic analysis.

Environmentalists, on the other hand, prefer high standards across the board. As in the preceding example, the interesting comparisons arise in the low/high cells. For these two cells, both environmentalists and industry exhibit the same direction of preference. Stricter standards for new sources are preferred by both groups to stricter standards for old sources. Industry adopts this preference ranking because lower costs to existing firms are more important than lower costs for new firms. Environmentalists adopt this ranking because they take a long-term outlook on environmental quality and assume that ultimately, environmental quality will be improved by having stricter standards for new firms.

The legislator is expected to balance the concerns of industry and environmentalists in a way that maximizes his net benefit. The choice of a particular cell by the legislator will depend on his utility function. Assuming that both environmentalists and industry have an important effect on this function, it is reasonable to expect that the choice reduces to the low/high cells, since the low/low and high/high are the least preferred alternatives of one group. But if the choice reduces to the low/high cells, the choice is relatively simple for the legislator. The cell with a high standard for new sources and a low standard for existing sources dominates its competitor for both interest groups, and consequently will be selected.[40]

4.3.2. The Movement Away From Standards

The use of this basic framework can be formalized and used to derive testable predictions. One of the areas that has received very little attention until recently is the emergence of incentive-based mechanisms to address environmental problems. Theories on the political feasibility of these mechanisms and the likely form these

---

[40] This example should not be taken too literally, but is useful for illustrative purposes. As a counterexample, consider a case in which emissions from existing sources were unregulated, but new sources were not allowed to emit any pollutants. In this case, no new sources would be built (assuming all sources pollute at some positive level). This is clearly not a case that either industry or environmentalists would view as an attractive bargain. I am indebted to my colleague. Gordon Hester, for suggesting this example. It raises an important point about the nature of the policies that are selected for consideration in Table V. It is reasonable to assume that these policies are undominated with respect to other feasible policies that would be placed in the same cell. This would eliminate the problem presented by the counterexample.

mechanisms will take are just beginning to emerge. The early work of Buchanan and Tullock (1975) gave rise to a steady stream of research on explaining the choice of standards. The instrument against which standards were most frequently judged were emissions taxes in their pure form. As several scholars have noted, emission fees are rarely implemented in ways even remotely resembling their pure form (see, e.g., Brown and Johnson, 1984). Consequently this instrument choice comparison may not be terribly revealing.

To develop a more realistic theory of instrument choice, it is necessary to explore how actual instruments behave in practice. The actual performance of incentive-based mechanisms varies widely. For example, a market in lead rights for controlling lead levels in gasoline has performed quite well in terms of efficiency, while a market for controlling emissions from air pollutants has not performed that well (Hahn and Hester, 1987a). Is it possible to account for such differences in performance, and if so, how? From a theoretical point of view, it is possible to ascribe these differences to several factors.

Suppose that industry and environmentalists have preferences over both the nature of instruments used and the overall level of environmental quality. Let M be a variable that characterizes the nature of instruments, and let Q represent the level of environmental quality. Environmental quality is relatively easy to measure, but the nature of instruments needs to be defined. In this case, M represents a single dimension that denotes the degree to which a system is "market" oriented. When $M = 0$, this corresponds to the case of conventional source-specific standards. When $M = 1$, this corresponds to a "pure" marketable permits approach. Values of M falling between 0 and 1 represent varying "degrees" of markets. This may seem like a peculiar concept in that either markets exist or they don't. However, markets are frequently governed by very different rules of exchange, and this variable attempts to capture the extent to which trading is restricted. For example, the market for lead rights would be associated with a value of M close to 1, while the markets for controlling air pollutant emissions would be associated with a value of M much closer to 0. As M increases, the efficiency of the instrument, measured in terms of aggregate reductions in cost savings, is presumed to increase.

The preferences of industry and environmentalists are given by the functions $I(M, Q)$ and $E(M, Q)$, respectively. The problem facing the regulator is to maximize his utility, which is assumed to be a linear combination of the preferences of environmentalists and industry.[41] Thus, the regulator will choose M and Q to

$$\underset{M,Q}{\text{Maximize }} a\, I(M, Q) + (1 - a)E(M, Q). \qquad (4.1)$$

In this problem, and in all subsequent variations of this problem, a is a weighting parameter which is assumed to vary between 0 and 1. The preferences of industry receive a high weight when a is close to 1.

The regulator's choice typically will be constrained by the requirement that the choice of M and Q be acceptable to both interest groups. Acceptability can be determined by whether the new policy is at least as good as the *status quo* for both groups. This requirement could easily be added to the formal constraint set. It is suppressed here in the interest of simplicity.

Assuming the function is differentiable, the first order conditions for an interior maximum are:

$$aI_1 + (1 - a)E_1 = 0$$

and

$$aI_2 + (1 - a)E_2 = 0,$$

where the subscripts on the I and E variables denote partial derivatives with respect to the arguments of the functions. For example, I_1 denotes $\partial I/\partial M$. The first order conditions state that a weighted sum of the marginal utilities will be 0.

Up to this point, nothing has been assumed about the precise form of the preferences of industry and environmentalists other than they are differentiable. To understand how M and Q are affected by changes in exogenous parameters, such as a, it is necessary to specify the nature of interest group preferences. In this, and all cases that follow, both industry and environmental

---

[41] In the formal analysis that follows, the word "regulator" will be used. However, it should be understood that the regulator could be a politician or a bureaucrat at any level of government.

preferences are assumed to be "well-behaved." In particular preferences are assumed to be representable by strictly concave functions that are twice differentiable. This assumption is made in the interest of simplicity, and because it is plausible for the situations represented here. Strictly concave preferences for industry and environmentalists imply that the regulator's maximization problem, which is a linear combination of these preferences, is also strictly concave.[42]

All that remains to be specified is the exact form of the utility functions. These will vary across the different cases presented here. For this particular case, industry and environmentalist preferences are characterized by the following set of partial derivatives:

$$I_1 > 0, \quad I_2 < 0, \quad I_{11} < 0, \quad I_{22} < 0, \quad I_{12} \geq 0,$$

and

$$E_1 < 0, \quad E_2 > 0, \quad E_{11} < 0, \quad E_{22} < 0 \quad E_{12} \geq 0.$$

Industry is assumed to prefer a more market-oriented alternative because it saves money.[43] However, there are decreasing returns to further movements in this direction. Industry prefers lower environmental standards, but again there are diminishing returns. Environmentalists, on the other hand, are distrustful of market alternatives and prefer the current standard-based approach. As the market orientation is lowered (M decreases), the marginal gain from a unit decrease in M is lower. Unlike industry, environmentalists prefer higher levels of environmental quality, but this is also subject to diminishing returns. These assumptions are farily standard. A critical assumption relates to the cross-partial derivative of both of these functions. In this case, both are assumed to be nonnegative. For industry, this says that as the environmental

---

[42] Strict concavity need not be assumed to obtain most of the results. However, concavity and differentiability are needed.

[43] This will certainly be true in the case of identical firms. However, when there are important cost differences among firms, this preference will not hold. Also, firm attitudes will be dramatically affected by the proposed distribution of property rights (Hahn and McGartland, 1988). When property rights are grandfathered, most firms could be expected to prefer a more market-oriented scheme since making the right tradable enhances its value. Nonetheless, there may be other strategic considerations, such as the reduction in barriers to entry, that induce existing firms to oppose a move to more market-oriented schemes.

quality standard increases, the marginal utility from using a market also increases. This results from the fact that higher levels of environmental quality are associated with greater gains from trade.[44] For environmentalists, an increase in the environmental standard is assumed not to decrease the attractiveness of markets. This assumption is certainly debatable. Thus, it will be necessary to discuss the implications of relaxing it.

Given these assumptions, it is possible to examine how M and Q will respond to changes in industry influence. This will provide insights into the conditions under which markets are likely to emerge. The basic result is given in Proposition 1.[45]

PROPOSITION 1: *An increase in industry influence will increase the market orientation of the instrument and reduce the level of environmental quality that is selected.*

There is a simple intuition behind this result.[46] As industry influence increases, environmentalist influence decreases. Thus, we will tend to observe more of what industry likes, and less of what environmentalists like. In this case, industry is assumed to like market-oriented alternatives and lower levels of environmental quality, since both can result in higher industry profits.

If the cross partial derivatives are negative or of unknown sign, then Proposition 1 does not hold. However, it is still possible to say something about the effects of a shift in the relative importance of environmentalist and industry preferences. This is summarized in Proposition 2.

PROPOSITION 2: *If preferences are well-behaved, an increase in industry influence will result either in a decrease in environmental quality and/or an increase in the market orientation of the instrument.*

---

[44] This assumption is consistent with most simulation studies I have seen over the region of interest (e.g., see Hahn and Noll, 1982a). Note, however, that as the overall level of emissions approaches 0, then industry will probably feel differently. To see this consider the extreme case where the emission limits are 0. Then all firms have the same emission standard, and a market adds no flexibility in this case.

[45] All proofs are provided in the appendix.

[46] The propositions examining a change in relative influence are stated in terms of the effects of a relative increase in the influence of industry. The effects of a relative increase in the influence of environmentalists are just the opposite of the effects of a relative increase in industry influence.

Propositions 1 and 2 provide insights into tradeoffs that are made among environmental quality and market orientation of the instrument. To understand the nature of these insights, it is instructive to explore what could induce changes in a, the variable that determines the relative influence of the two interest groups. The extent of influence is affected by several factors, including the attitudes of politicians, the cost of accessing key decision makers, and the resources available to different interest groups. In addition changes in institutions will also dramatically affect the relative influence of interest groups. By drawing on examples where preferences and institutions remain relatively stable, while some of these other factors vary, the theory presented above can be subjected to empirical verification.

Propositions 1 and 2 are derived under the assumption that the effects of market-based alternatives are certain.[47] One of the key issues that affects the emergence of markets and the type of market selected is uncertainty over the instrument's performance. Generally, lower uncertainty will result in greater use of market-based alternatives. To see how uncertainty can be included in the analysis, consider the following extension of the basic maximization problem given by (4.1):

$$\text{Maximize}_{M,Q} \; a(I(M, Q) + \Phi(M)) + (1 - a)(E(M, Q) + \Psi(M)). \quad (4.2)$$

The functions $\Phi(M)$ and $\Psi(M)$ explicitly introduce a special term representing the effects of uncertainty on the utility derived by industry and environmentalists. For simplicity, the utility associated with the uncertainty of the instrument is captured in a separate function for both industry and environmentalists. The effect of uncertainty can be expected to vary with the degree of market orientation. Generally, the greater the departure from the *status quo* (i.e., $M = 0$), the greater the "costs" of uncertainty. This idea can be expressed by assuming that both Φ and Ψ are strictly concave. In particular, $\Phi_1 < 0$, $\Phi_{11} < 0$, $\Psi_1 < 0$, $\Psi_{11} < 0$. Under the case of pure standards, uncertainty is assumed to have no negative effects (i.e., $\Phi(0) = 0$ and $\Psi(0) = 0$).

There are two questions that this formulation of the problem can

---

[47] Alternatively, the uncertainty is already factored into the preference structure.

address. The first is the effect of the uncertainty on the adoption of markets and the second is the effect that uncertainty will have on the type of markets that are adopted.

Consider the effect of uncertainty on the adoption of markets. The addition of uncertainty lowers the value of the objective function. Consequently, it makes the *status quo* (where $M = 0$) look that much more attractive. It is possible to go further than this when we recognize that the functions characterizing uncertainty may change wth time. In particular, as experience is gained with market-based alternatives, one would expect that the uncertainty associated with their use would decline. This has an important implication, which is stated in Proposition 3.

PROPOSITION 3:   *As uncertainty associated with any level of market orientation declines, it becomes more likely that a market-based approach will be chosen.*

This result suggests that there is a type of "demonstration effect" associated with the use of market alternatives. As they are used more, uncertainty about their use is reduced. Consequently, they are more likely to be used in the future. This is a prediction about the future use of market-based approaches. It needs to be tempered by the fact that, in reality, the increased use of these mechanisms may not only affect the level of uncertainty associated with their use, but can also affect attitudes towards their use. Thus, for example, if these mechanisms tend to perform well in terms of lowering costs and improving environmental quality, their use can be expected to increase. However, if their application is associated, rightly or wrongly, with a bad outcome, such as increased levels of noncompliance or lower environmental quality, their application could actually decrease. Thus far, actual experience with market-based approaches has resulted in substantial cost savings and very little change in environmental quality. To the extent that this is perceived by key interest groups, it should foster the use of these mechanisms.

The preceding analysis shows that the introduction of uncertainty can decrease the likelihood that a market-oriented alternative will be chosen by making it less likely that firms will move from the *status quo*. However, suppose there is an interior solution to the maximization problem defined by (4.1) that dominates the *status quo.* How will the introduction of uncertainty affect this solution? One would expect that the value of *M* that was optimal for the original problem exceeds the value of *M* that was optimal for the problem in which uncertainty is included. However, this is not true in general, and will hinge on the cross-partials associated with *I* and *E*. Preferences will be said to be "independent" for the special case in which the cross-partial derivatives are zero. For this special case, it is possible to derive the following result:

PROPOSITION 4: *If preferences are independent, then the optimal level of M will decline upon introducing uncertainty.*

This result is interesting because it suggests that the introduction of uncertainty need not lead to a decline in *M* in all situations. This is somewhat surprising, but again reflects the fact that the tradeoffs that industry and environmentalists are willing to make among market orientation and environmental quality are crucial to understanding the type of instrument that is selected.

While the preceding analysis has been cast in terms of market orientation, the *M* variable can be interpreted somewhat more broadly. It is possible to use this approach to differentiate between various types of standards. Standards that allow firms greater flexibility in meeting emission limits could be assigned a higher value of *M*. For example, performance standards typically require that firms meet a prescribed emission limit, but they allow the firm to choose the abatement technology for meeting this limit. In contrast, technology-based standards dictate the precise form of technology to be used. The point is that the theory developed here is also relevant for explaining the broad menu of standards that are currently in use.

A similar theory can be used to explain the use of emission fees. Emission fees are primarily used as a means of raising *revenues*. These revenues are almost always earmarked for improving specific environmental problems associated with the pollutants that are subject to the fee. Only in a few applications have emission fees been shown to have a marked incentive effect.[48] The choice about the type of fees that are selected can be succinctly modeled by

---

[48] See Chapter 3 for a review of the literature on fees and an assessment of their performance.

assuming that interest groups have well-defined preferences over the size of fees, F, and how they are used, U. Higher levels of fees are associated with higher values of F. Greater earmarking for specific environmental improvements is associated with higher values of U. This yields the following maximization problem for the regulator.

$$\underset{F,U}{\text{Maximize}}\ a\,I(F,\,U) + (1-a)E(F,\,U). \qquad (4.3)$$

To understand how choices change with different weightings, the preferences need to be defined. Industry and environmentalist preferences are characterized by the following set of partial derivatives:

$$I_1 < 0, \quad I_2 > 0, \quad I_{11} < 0, \quad I_{22} < 0,$$

and

$$E_1 > 0, \quad E_2 > 0, \quad E_{11} < 0, \quad E_{22} < 0.$$

Industry prefers lower fees to higher fees because fees represent an extra cost of doing business. Environmentalists prefer higher fees becuase they will reduce pollution either directly, through their impact on firm decisions to pollute, or indirectly, through their impact on expenditures aimed at reducing pollution. Both groups prefer the earmarking of fees—industry, because it increases the credit they can claim for reducing pollution, and environmentalists because they are in favor of activities that promote environmental quality. Both functions reveal diminishing marginal returns in F and U. The cross partial effects are more difficult to predict, and again are key to predicting the effect of shifts in influence on the level of fees and the degree of earmarking. The results of a shift in the relative influence of industry are summarized in Proposition 5.

PROPOSITION 5: *An increase in the relative influence of industry will result in a decrease in fees if preferences are independent. An increase in the relative influence of industry will result in a decrease in fees and no change or an increase in earmarking if the cross partials are non-negative and the marginal utility of earmarking for environmentalists does not exceed the marginal utility of earmarking for industry.*

If preferences are well-behaved, an increase in industry will result in a decrease in fees and/or an increase in earmarking.

Part of this proposition conforms to intuition. As greater influence is given to industry preferences, fees are reduced, since industry prefers lower fees and environmentalists prefer higher fees. The situation with earmarking is less clear, since both groups prefer higher levels of earmarking. There are a variety of fees that are currently in use for activities ranging from aircraft noise to hazardous waste disposal. These fees exhibit wide variations in their effects. Even within categories of pollutants, fees vary widely across industries and jurisdictions. This model argues that part of this variation is attributable to the relative influence of industry and environmentalists.

It is worth noting that the structure of this model suggests that earmarking is a very stable feature of the political landscape. Economists have often criticized earmarking because it restricts the flexibility that the government has in allocating its budget. This argument is not likely to hold the day when strong interest groups are interested in claiming credit for state expenditures, especially when the state chooses to impose a special tax on specific industries that are influential.

The two models of fees and market-based activities look at different aspects of choice within these two classes of activities. This provides insights on the nature of instruments that are likely to be chosen within these classes. Frequently, however, both classes of instruments may be considered in the actual application of instruments. An example will illustrate the nature of choices that are involved.[49] The state of Wisconsin is in the process of devising a plan to help address the problem of meeting the ozone problem in the southeast portion of the state. The state has generated a surplus of emissions rights for hydrocarbons, one of the major contributors to the formation of ozone. The problem confronting the regulators and legislators is how to allocate this surplus. After considering charging a fee that reflects the marginal cost of a permit, or trying to create a market in permits, the state has opted for a regulatory strategy based on first-come first-serve, with a nominal one-time fee

---

[49] This section draws heavily on analysis contained in Hahn (1987b).

TABLE VI
Instrument Choice When Jobs Matter

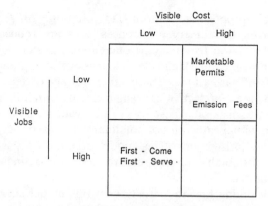

Visible   Cost

	Low	High
Low		Marketable Permits Emission Fees
Visible Jobs **High**	First - Come First - Serve ·	

attached to the use of the permits.[50] The decision to adopt this approach can be understood in terms of political costs and benefits. Politicians in this region are very interested in promoting jobs, and particularly jobs that are quite visible. At the same time they are interested in promoting regulatory systems that appear to have a low cost to industry.

The choices open to regulators and politicians are summarized in Table VI. The three basic instruments are presented in the table. The policy which was selected is denoted as first-come first-serve, since this is the most salient feature of the policy. The costs of this policy are not readily apparent to the population at large. They include the cost of generating surplus emissions reductions, and the efficiency costs associated with the fact that the mechanism does not induce firms to search for more efficient approaches to pollution abatement. While the policy does not do well when measured in terms of efficiency, it does quite well on the dimension of visible job creation. It is designed to help accommodate "blockbuster" projects that would bring large numbers of jobs to a depressed economic region.

---

[50] Interestingly, first-come first-serve is used quite frequently in the initial allocation of many types of property rights. For example, businesses locating in relatively clean areas are typically allowed to locate there without purchasing emission rights until the surplus in emission rights is exhausted. These businesses do have to comply with existing state and federal standards.

In contrast, marketable permits and an emissions fee that is based on the marginal cost of abatement share the opposite characteristics. The costs of these policies are quite visible in the sense that these costs fall directly on industry. Industry can readily identify these costs in terms of tax expenditures or expenditures for permits. The efficiency gains associated with these policies tend to be more diffuse. Moreover, relative to the case of first-come first-serve, job creation is not as visible. No state entity has to be consulted before using these pollution rights, and there is no explicit need to justify the use of these rights on the basis of the number of jobs created. To the extent that direct job creation occurs, much of it may occur through relatively small changes in the use of inputs such as pollution and labor.

In short, these two policies are associated with highly visible costs and jobs with low visibility. This is just what politicians don't want. On the other hand, the first-come first-serve policy has the desired characteristics. It appears to be relatively low in cost and promotes the highly visible forms of job creation. The lesson to be learned from this example is that there may be strong forces working at the local level to impede the development of incentive-based alternatives. However, this is not universally true. Indeed, in cases where abatement costs are relatively high, and jobs are relatively less important in the political calculus, the appearance of market-based alternatives is more likely.[51]

4.3.3. Broader Patterns in Environmental Regulation
While the preceding arguments highlight the use of various instruments in meeting environmental quality objectives, much of the struggle in the environmental arena concerns the choice of an objective itself. Often, objectives are selected that are not met. For example, Congress once called for eliminating all discharges into navigable waterways by 1985. In addition, Congress has repeatedly mandated standards for air pollutants that were not met in the specified time frame. The Clean Air Act was amended in 1977 because it did not meet certain goals such as the standard for ozone. In 1989, this same act will probably be amended again because the targets were not met. This consistent pattern of falling short of the

---

[51] Los Angeles is a case in point.

stated targets suggests that legislators may not really intend for targets to be met in the specified time frame. It is worthwhile exploring the possibility that a key motivation for adopting such goals is rooted in their symbolic value.

The general importance of symbols in politics is well-known (Edelman, 1964). They provide benefits for politicians who are interested in mobilizing specific constituencies. They can also provide benefits to those constituencies as well. These general insights still leave some important unanswered questions pertaining to the use of symbols in environmental regulation. For example, what motivates the consistent pattern of behavior where targets are frequently not met in the specified time frame? Alternatively, why are laws and regulations passed that appear to be quite stringent, but the incentive to comply with these laws based on existing monitoring and enforcement capabilities is quite low?

A partial answer to these questions can be found by examining the payoff to different interest groups. Environmentalists may want a high symbolic value for environmental quality for several reasons. First, it may signify a long-term commitment to a goal. Thus, standards are set either at the limits of technological feasibility or beyond the realm of what is currently feasible. Implementation then proceeds at a much slower pace, and only a small fraction of resources are allocated to monitoring and enforcement activities. Another explanation is that symbols can help influence the pre-ference structure and values of individuals. Thus, environmentalists may want future generations to be imbued with an environmentalist ethic. Industry, on the other hand, may be opposed to symbols, in the sense that symbols can affect actual targets over the long term; and higher actual targets are frequently associated with higher costs.[52]

These contrasting attitudes towards symbols can be illustrated in a simple model. Consider a two period model where industry and environmentalists can negotiate over the level of environmental

---

[52] A third possibility not captured here is that the symbol itself may have intrinsic value, separate from its effects on preferences or physical outputs. This possibility is not considered explicitly in the formal analysis, though it could easily be incorporated.

TABLE VII
An Example of the Feedback
Effect Between the Stated Target
and the Actual Level

Period 1		Period 2
$[Q^1, S]$	$\longrightarrow$	$[S, S]$
$[Q^2, Q^2]$	$\longrightarrow$	$[Q^2, Q^2]$

quality. They have two variables over which they can negotiate—the actual level of environmental quality level, Q, and the target level, S. Suppose, for simplicity, that the symbolic target in period 1 becomes the actual level in period 2. The regulator selects values for Q and S in the first period. These values then determine the actual value for Q in the second period. The value of S in the second period is assumed to be equal to the actual level achieved in the second period. Table VII elaborates on this possibility. Each ordered pair gives the value of Q and S respectively. Two cases are examined in the table. The point is to compare a case where actual environmental quality does not vary across periods against one in which environmental quality increases with time. In the first case, the regulator sets $Q = Q^1$ and also picks a value of $S > Q^1$. This results in a second period actual value of S. In the second case, the regulator chooses a value for $Q = Q^2$ and $S = Q^2$, so the actual environmental quality does not change in the second period. Now suppose that $S > Q^2 > Q^1$. Which option does the regulator choose? Again, this depends on the underlying preferences of industry and environmentalists. However, one could easily imagine values for the parameters that result in choosing the first strategy. A relatively low actual level in period 1 would result in lower costs for industry. These costs would have to be weighed against the higher costs in period 2. The low environmental quality in period 1 might not be of major concern to environmentalists if the ultimate level that was achieved was high. This is consistent with the view that environmentalists may implicitly choose to heavily discount the present.

It is possible to integrate the basic tradeoff suggested by Table VII into a more general model. Suppose industry and environmentalists have preferences over environmental quality, Q, and the target level of environmental quality, S. This yields the now familiar maximization which the regulator must solve:

$$\underset{Q,S}{\text{Maximize}}\; a\, I(Q, S) + (1 - a)E(Q, S). \tag{4.4}$$

Industry and environmentalist preferences are characterized by the following set of partial derivatives:

$$I_1 < 0, \quad I_2 < 0, \quad I_{11} < 0, \quad I_{22} < 0,$$

and

$$E_1 > 0, \quad E_2 > 0, \quad E_{11} < 0, \quad E_{22} < 0.$$

Industry prefers lower environmental quality and a lower target level of environmental quality since both of these are presumed to increase costs. In contrast, environmentalists prefer higher environmental quality and a higher target level of environmental quality because both of these increase actual environmental quality. Moreover, the symbol may have some intrinsic value. The effect of the cross-partial derivatives is not easy to predict. A shift towards greater industry influence will tend to reduce both environmental quality and/or the prescribed target under the conditions specified by the following proposition.

PROPOSITION 6: *If preferences are independent, then an increase in industry influence decreases the actual level of environmental quality along with the target level of environmental quality. If preferences are well-behaved, then an increase in industry influence will either decrease actual environmental quality and/or the target level of environmental quality.*

This proposition suggests that we should expect to see larger differences between prescribed targets and actual levels where environmentalists wield greater influence.

A problem which has a very similar structure to this problem has been examined in great detail by Mendeloff (1988). The basic thread of Mendeloff's argument is that relatively few chemicals and hazards get regulated, but those that do get regulated very stringently. Mendeloff draws examples from regulation of chemical exposures in the workplace and toxic substances regulation to demonstrate this observation. He casts the underlying dynamics of this problem in terms of a two interest group model consisting of labor and industry (Mendeloff, 1988, Figure 9–2). While from an efficiency point of view it may be desirable to have more moderate regulation of a wider group of chemicals, this conflicts with the preferences of labor. Moreover, in the current political environment, it is difficult to develop a system of payoffs that would entice labor to support this policy.

It is possible to understand the problem examined by Mendeloff in terms of the maximization given by (4.4) by redefining the variables. Let Q represent the breadth of regulation, which could correspond to the number of chemicals of a particular type that are regulated; A higher value of Q would mean that more chemicals get regulated. Let S represent the stringency of regulation. A higher value of S is associated with more stringent regulations on particular chemicals. The preferences of environmentalists and industry are again diametrically opposed, and can be represented by those given in problem (4.4). This gives rise to the following corollary, which directly follows from Proposition 6:

COROLLARY 1: *If preferences are independent, then an increase in industry influence decreases the breadth and stringency of regulation. If preferences are well-behaved, then an increase in industry influence will either decrease breadth and/or the stringency of regulation.*

This corollary has testable implications. It suggests that relatively few chemicals will be regulated, and that regulations will tend to be less stringent, when industry wields greater influence.

Of course, the nature of the influence that industry wields can matter. During the first term of the Reagan Administration, industry appears to have transcended the bounds of good taste. When Administrator Burford was running the Environmental Protection Agency, Congress became incensed at what it viewed as regulatory mismanagement (Florio, 1986). The response to this recalcitrance was to introduce "hammer" legislation that required EPA to act by certain deadlines. In the event that EPA did not act, stringent regulations would be imposed on industry for controlling hazardous waste and groundwater contamination (Hahn, 1986a). This is an example where the increase in industry influence runs counter to the prediction of the theory. The theory would not appear to hold in situations where Congress perceives major problems in the execution of its mandate. In such cases the lawmakers tend to adopt highly symbolic solutions to problems that are frequently quite expensive and not well-thought out.

Up to this point, the focus in this section has been on explaining specific aspects of instrument choice in environmental policy. One of the important themes that arises in the implementation of virtually all environmental policies is that they are multifaceted. Indeed, almost every incentive-based system involves the use of several instruments. For example, some type of standard lies at the heart of most environmental regulatory systems, even those which feature fees and marketable permits. Moreover, a system of monitoring and enforcement is required to ensure that most systems will achieve some degree of compliance. Given the pervasive use of multiple instruments, there is a need to explain this phenomenon. Perhaps, the simplest explanation for this phenomenon is that the implementation of most environmental policies requires several steps. These steps include defining the general problem, providing specific guidelines to firms and ensuring that firms will meet these guidelines (e.g., see Bohm and Russell, 1985). It is unreasonable to think that a single instrument is likely to be suited to the myriad of tasks involved in implementing an environmental policy. This is true regardless of whether a philosopher-king implements the environmental policy or the policy is implemented by mere mortals governed by political forces. In both cases, the use of multiple instruments will tend to be the rule rather than the exception. A simple way of thinking about multiple instruments is by adopting the conventional paradigm used in environmental economics. Imagine that there are benefits and costs associated with using different configurations of instruments. The regulator or legislator chooses a set of instruments to maximize a prescribed political objective function. This will involve trading off between the cost of using additional instruments and their marginal benefits (e.g., see Hahn, 1986). Thus, the problem of using multiple instruments can be conveniently described in terms of the conventional maximization calculus used here.

The preceding analysis has couched the instrument choice problem in terms of a regulator or legislator maximizing a political support function or balancing the competing claims of industry and environmentalists. This paradigm was helpful in three general areas.

First, it added to our understanding of the standard-setting process by providing insights on how standards are likely to be implemented. Second, it was useful in identifying conditions under which incentive-based instruments will emerge, and identifying the type of incentive-based instruments that are likely to be adopted. Finally, it was useful in explaining some broader patterns in environmental regulation. Examples included the breadth and stringency of regulation, as well as the tendency to adopt symbolic goals.

4.4. Modeling Issues

The theory presented here has focused on some central aspects of environmental policy. It also has left out some very important parts of the problem. For example, the issue of subsidies was not explicitly addressed. The political motivation underlying subsidies is fairly well understood. Subsidies enable politicians and bureaucrats to take credit for supplying specific benefits to their constituencies. A good example from the field of environmental regulation is the huge subsidy for municipal sewage waste treatment plants (Arnold, 1979). While it is easy to understand the general use of subsidies in the political process, relatively little is known about the determination of subsidy levels and the geographic distribution of subsidies. Becker (1983) suggests a model for income transfer that sheds some light on the features of subsidies, but the model is not suited to answering questions about their observed levels. The problem of geographic distribution of political "pork" such as subsidies is still the subject of heated debate (Ferejohn, 1974; Arnold, 1979). These unresolved questions about subsidies point out an important limitation of the modeling approach that has been adopted in this paper. It does not take advantage of many of the important *institutional* features that shape regulatory policy (Noll, 1983; Weingast, 1981). Nonetheless, it can be helpful in addressing certain parts of the subsidy issue. In the case of municipal waste treatment plants, for example, there has been a marked tendency on the part of the federal government to provide major subsidies for capital expenditures, but to require states to shoulder the operation and maintenance costs. This can be modeled in terms of the payoff to a single congressman, who gets most of his credit up front, with the initial ground breaking ceremony for the plant.

Providing a more comprehensive theory of environmental policy design will require a careful look at the institutions that shape this design. The importance of the organization of the Congress and comparable legislative institutions in other countries has been pointed out by several scholars (e.g., see Fenno, 1973; Noll, 1983). Other scholars have tended to focus on the importance of the bureaucracy and the courts (Melnick, 1983; Wilson, 1980). Relatively few studies have been done that examine how these organizations have helped shape the type of instruments examined in this section (Meidinger, 1985). Hahn and Hester (1986) have argued that forces within the bureaucracy had a major impact on the development of the "emissions trading" policies by the Environmental Protection Agency. In the case of market-based reforms, there appears to be an important role played by both academics and "bureaucratic entrepreneurs" who are trying to take credit for new ideas.

Liroff (1986) has chronicled some of the differences in views that exist in different parts of EPA. There is an important difference between parts of the agency dedicated to implementing programs (the "program offices") and the part of the agency dedicated to evaluating policy (the "policy office"). Program offices are interested primarily in implementing their regulatory mandate. Crudely speaking, they get evaluated on producing regulations. The policy office, on the other hand, does not have a specific regulatory mandate. Members of this office get evaluated on their attempts to produce more efficient regulation. Not surprisingly, the impetus for both major marketable permit programs has tended to come from the policy office and other parts of the government interested in promoting more efficient forms of environmental regulation (Hahn and Hester, 1987a). Nonetheless, both the program office for managing air pollution and the policy office have had major impacts on actual rules regarding the trading of rights to emit air pollutants. This suggests that bureaucratic incentives can and do play an important role in affecting the emergence and design of policies.[53]

At the same time, it would be misleading to imply that the bureaucracy shapes policies completely independently of either the

---

[53] For a more extended discussion about the role of bureaucracy in shaping outcomes, see Hahn and McGartland (1988).

Congress or key interest groups. Indeed, the bureaucracy is constantly trying to gather support for its actions from all of these groups. What the empirical analysis does suggest is that the bureaucracy is not necessarily best viewed as a passive agent that carries out the wishes of Congress.

The relationship between the bureaucracy and other political institutions may be critical for determining policy outcomes. Several scholars have commented on the differences in the importance and style of bureaucracies across countries (e.g., see Brickman, Jasanoff and Ilgen, 1985). While it is clear that bureaucracies differ, it is less clear how this substantively affects policy outcomes. However, bureaucracies that are seen as agents of the ruling party (when there is one), may develop quite different policies than bureaucracies that have to balance the interests of an executive and legislative branch dominated by different parties. The reason is that politicians will face different payoffs in the two cases. In the case where legislators and the executive are dominated by different parties, legislators may try to use the bureaucracy in ways that make the executive look bad.

There is another very important sense in which the bureaucracy matters in considering problems in instrument choice. The preceding formal analysis was built on the assumption that certain tradeoffs could be made among different dimensions of policy. Trading, and the nature of trading, is likely to be constrained by the design of political institutions. This includes the design of legislative institutions, the courts and bureaucracies. In the case of EPA, for example, issues in monitoring and enforcement are carried out largely independently of standard-setting. This means that there is little opportunity to effect trades on these issues at the bureaucratic level. Thus, the principal opportunity for "trading" in this area would be at the legislative level. This example reveals that a careful analysis of organizational design can provide insights into the potential for bargaining as well as the likely arena in which bargaining will take place.

Another important issue related to the study of bureaucracies is that of delegation. There has been a great deal of study of legislative delegation of authority (Aranson, Gellhorn, and Robinson, 1982; Fiorina, 1982). However, it is important to recognize that bureaucracies have choices in what they delegate to other bureaucracies. For example, EPA in its recent revision of its trading policies made easier for the state environmental agencies to develop programs with less federal oversight. Moreover, there is evidence that decreased levels of oversight are associated with increases in the efficiency of emissions trading (Hahn and Hester, 1986). It would be useful to have a theory of why bureaucracies delegate in some instances and not others, and the expected effects of delegation in terms of efficiency and equity. Such a theory could be built on existing legislative theories of delegation (e.g., see McCubbins, 1985).

The preceding discussion reveals that instrument choice, like many other political decisions, is driven by a wide array of interest groups both in and outside of government. Yet the formal modeling approach used here focuses on a single, or representative, decision maker. This is obviously a gross oversimplification. Nonetheless, it is useful for helping to understand some of the broad outlines of environmental policy. Moreover, more realistic attempts to include the interrelationships among key groups influencing instrument choice decisions quickly leads to an analytical quagmire. In the past, scholars have attempted to deal with this problem by modeling salient aspects of the institutional process that are analytically tractable, such as the committee structure in Congress. This institutionalist approach is quite useful when the institution being modeled is the driving force behind the problem. However, it can also be quite misleading if the institution represents only one of many key actors in the decision making process. For example, in many of the cases examined here, the bureaucracy was seen as the prime mover or a major participant in many of the key decisions. Thus, detailed modeling of institutions other than the bureaucracy may lead to only marginal gains in understanding these decisions. The point is not that institutional analysis is not needed, but rather that great care should be exercised in choosing the appropriate institutional focus.

This entire section has been devoted to constructing a more complete theory of instrument choice. This theory has been built using two important assumptions: First, that different policy instruments can be distinguished on the basis of their distributive implications; and second that the set of available instruments can be specified. Given the nature of the theory, it cannot be expected to distinguish between instruments that have similar distributive consequences. This observation points to an important limitation of existing theories of instrument choice. To the extent that instrument choice is motivated by reasons unrelated to distributional concerns, the theory does not contribute to our understanding. However, there is a deeper problem with theories of instrument choice that relates to the second building block on which the theory rests. The assumption that the feasible space of instruments can be specified is problematic. Certainly, instruments that are being used can be identified. Sometimes, it is also possible to identify instruments that were considered at some point in the decision process, but were not selected. However, defining the entire feasible space of instruments is virtually impossible. At best, we can hope to get a reasonable grasp of political constraints that place limitations on the choice of instruments.

This raises the question of how to judge a theory of instrument choice, and one on which surprisingly little has been written. Certainly, one would like a theory that predicts what instruments are likely to be chosen under different conditions. It would also be useful to know what instruments are not likely to be chosen. The real art in developing a theory of instrument choice enters in defining the choice set. Until recently, the choice set has been defined more by theory than by empirical realities. Thus, for example, Buchanan and Tullock (1975) choose to "explain" the choice of standards by choosing what, upon closer inspection, appears to be a very unlikely alternative. At a minimum, a theory of instrument choice should try to explain important characteristics of instruments that exist in the real world. To the extent possible, it should place instruments in the feasible space that are "reasonable" competitors to existing instruments. To rationalize the existence of an instrument by comparing it against an unlikely alternative is not terribly useful.[54]

Related to the issue of defining the feasible set is the vexing problem of defining precisely what is meant by an instrument. For practical reasons, it would be useful to define instruments that are

---

[54] Actually, there are situations in which such rationalizations may be useful in redirecting the energies of academics. For example, such an endeavor could help to convince members of the economics profession that wholesale adoption of marketable permits or emission fees in their pure form is highly unlikely.

measurable, and that are likely to have systematic effects on policy outputs. From a theoretical standpoint, instruments can only be distinguished on the basis of their distributional properties. Consequently, in theory, it is often possible to design standard and tax systems that are indistinguishable. "Different" instruments may have similar political payoffs, and therefore may not be different in terms of their theoretical properties. The point is that words like "standard" and "tax" have meaning in terms of the theory only to the extent that they imply a particular distributional outcome. While it is useful to retain the general terminology, I also believe it is important to be very clear about the precise nature of the comparisons. For example, Buchanan and Tullock (1975) do not show why industry has a marked preference for standards over taxes. They show why industry has a marked preference for a very specific standard over a very specific tax.

In addition to being careful about instrument definition within a particular class, such as standards or taxes, problems can arise in making distinctions among classes. For example, to what extent should instruments such as standards and taxes be distinguished from monitoring and enforcement mechanisms? Ideally, it would be nice to merge many of these classes to do more global comparisons of instruments. Unfortunately, in many cases, the problems become analytically intractable. Thus, it is necessary to break the problem of instrument choice into manageable cases. The advantage of doing this is that it enables us to clearly understand the political and economic forces acting on one particular part of the problem. The disadvantage is that the separation may be artificial. Monitoring and enforcement mechanisms are inextricably linked to the choice of using pricing and quantity approaches for regulating pollution. In constructing theories of instrument choice, it is important to be cognizant of how these linkages can affect the validity of the theory.

4.5. Conclusions and Areas for Future Research

This section has illustrated how formal models of instrument choice can help explain important elements of environmental regulation. While developed primarily to explain themes in environmental policy, many of the theories have broader applicability. For example, the theories of standard-setting apply to the general field of regulation. The basic framework using competing interest groups was helpful in explaining the process of standard-setting, the emergence of new regulatory approaches, and some broad patterns in environmental regulation.

The primary thrust of this section has been theoretical. Its principal purpose was to provide an alternative framework for thinking about instrument choice problems. The theories developed here are not easily tested for two reasons. The first relates to the problem of specifying utility functions. The second relates to identifying key decision makers. In many cases testable versions of these theories or related theories will require a more careful elaboration of the institutional environment in which decision making takes place. While these obstacles are formidable, they are not insurmountable. Indeed, I would argue that the most pressing need at this point is to assemble a reasonable data base from which to begin to judge different theories of instrument choice. The increased use of incentive-based approaches provides a unique opportunity to examine different theories. At this point, it is also necessary to develop a more extensive data base on conventional instruments such as standards and subisides. There is little systematic data on the use of different instruments and the political forces leading to their selection. While this task is not an easy one, case studies could begin to form the basis for constructing a useful data base.

From a practitioner's point of view, it would be helpful to know whether different instruments perform differently, and if so, the underlying reasons. The analysis in Section 3 revealed that there is substantial variation across instrument performance, even among instruments that are nominally in the same class. Identification of differences in performance will be useful for bureaucrats charged with designing programs. One key area where very little is known is how the division of responsibilities among federal, state and local regulators affects program performance. For example, it is often thought that local environmental agencies are more likely to be responsive to the demands of industry; however, evidence supporting this view is largely anecdotal.

In the analysis of the emergence of market-based incentives, the argument was made that perceptions of performance can matter a great deal for instrument choice. As the properties of instruments become more widely known, this can affect the attitudes of key interest groups, including the bureaucracy, towards their use. The widespread application of effluent fees in Europe represent a good place to examine the extent to which such "demonstration effects" are important.

The general effect of information on instrument choice requires more careful scrutiny. There are important similarities across developed countries in what is regulated (e.g., see Brickman, Jasanoff and Ilgen, 1985). This suggests that information transmission across countries is an important variable in explaining what things get regulated, and the type of regulations that are applied. More work on the timing of regulations and their similarities and differences could help clarify the role of information and networks in the regulatory process.

The theory of instrument choice is still in its infancy. There are many ways in which it could be extended. Earlier, the issue of why bureaucracies choose to delegate certain types of tasks was suggested as an area for study. A more general question that has intrigued economists relates to the relative efficiency of policy choices made by political institutions. At present, there is widespread agreement that there is no reason to presume that government policies will be efficient (Becker, 1983; Shepsle and Weingast, 1984). However, very little is known about the degree to which government policies are likely to deviate from an efficient solution (assuming such solutions can be defined) in specific instances.

This section has suggested one vehicle by which efficiency can enter into the choice of incentive-based instruments. However, this is an area that needs a great deal more elaboration. One important factor affecting the efficiency of various regulatory approaches is the ability to monitor and enforce sidepayments. For example, in the case of health and safety, it may not be possible to induce labor unions to agree to broader regulation that is less stringent because there is no way of effecting the necessary payoffs. In the case of environmental problems, there may be no way to get industry to agree to a broader scope for toxic substance policy because they have no assurances that the resulting regulations will not be draconian. Political constraints on law makers will impose substantial barriers towards moving to more efficient short-run policies.

For those scholars interested in fashioning more efficient policies, the challenge still remains to identify conditions under which such policies are likely to emerge. The models on this subject to date are very general and also lead to highly ambiguous results. Perhaps, it is necessary to trade off some generality in the interest of understanding the likely performance of specific policy arenas. Hopefully, this section represents a first step towards this end.

5. TOWARDS USABLE KNOWLEDGE: AN EXERCISE IN DESIGNING NEW REGULATORY APPROACHES

5.1. Introduction

The preceding two sections characterized the nature of existing incentive-based mechanisms. This section provides an application to a concrete problem that should have been decided by the time this book is published.[55] The fact that the issue may have been decided need not deter you from reading further. The basic problem is likely to arise in several different contexts, albeit in slightly different forms.

The primary objective of this section is to show how to apply some of the basic insights to the previous sections in an institutional context. This exercise in design will enable us to carefully examine some of the issues related to efficiency and feasibility that were raised in the previous sections. It will also show the importance of understanding the interrelationships between institutions in developing sensible alternatives for achieving environmental quality goals.

The particular application involves developing market-based approaches for the Clean Air Act. The objective will be to develop mechanisms that can result in achieving the ozone standard in areas that are characterized by high levels of pollution ("nonattainment areas"). Part 5.2 provides a brief definition of the ozone nonattainment problem in the United States. A series of alternatives are proposed and evaluated in Part 5.3. Part 5.4 highlights the key findings and discusses areas for future research.

---

[55] Given the propensity of Congress to defer difficult decisions, the decision point may be deferred.

5.2. Problem Statement

By the end of 1987, the entire U.S. was required to be in compliance with the 0.12 parts per million standard for ground level ozone (National Clean Air Coalition, 1985). If states or selected regions of the country were not in compliance, the U.S. Environmental Protection Agency could have applied a series of rather draconian sanctions. These sanctions included severely restricting federal funds for the highways, sewage treatment plants, and air quality programs, and imposing a ban on construction of major new sources (National Clean Air Coalition, 1985). The EPA Administrator, Lee Thomas, identified 76 regions of the country as nonattainment areas (Thomas, 1987).[56] While most of these areas were quite close to being in compliance with the law, a few, such as Los Angeles and Houston, have a long way to go.

Congress and the EPA faced a difficult dilemma. The deadline was approaching, the costs of meeting the deadline were unacceptable, and there was no obvious path to follow. The EPA Administrator characterized the situation this way:

... the public difference of opinion about EPA's future course of action mirrors the complexity of the ozone issue itself. There is no obvious course of action. There is no simple course of action. Any decision on an ozone strategy will have to balance a number of factors that are more or less important depending on your point of view.[57]

The problem of ozone nonattainment is not new. There is already a detailed infrastructure that has been created to help meet the ozone standard. Congress has grappled with this problem in the 1970 Clean Air Act Amendments and the 1977 Clean Air Act Amendments. It is useful to examine the basic design of this Act as a first step in developing reform proposals.[58] In order to comply with the national ozone standard of 0.12 parts per million, the Act requires that states submit a plan that has to be approved by the EPA. State Implementation Plans for ozone were required to be submitted in 1979 for ozone nonattainment areas. Areas that could

---

[56] See information presented in the first table of the testimony. Nonattainment areas are defined as those areas that do not meet the federally mandated national ambient standards.

[57] Thomas, 1987, pp. 17–18.

[58] See Stewart and Krier (1978) for a more complete account.

not demonstrate attainment by 1982 had to submit revisions in 1982 that would demonstrate attainment no later than 1987.

The plans generally consist of a series of control measures aimed primarily at limiting volatile organic compound (VOC) emissions from existing stationary sources. In addition, many areas have implemented emissions inspection and maintenance programs that try to reduce emissions of hydrocarbons from motor vehicles. Most of the implementation plans require specific reductions from individual sources or classes of sources. In many cases, even the nature of the technology is specified. This is the cornerstone of the so-called "command and control" approach. In some instances, other alternatives have been explored. For example, the EPA has attempted to promote the trading of "emission reduction credits" as a way of providing firms with greater flexibility in meeting emissions objectives. These credits are limited forms of property rights that can be created when firms reduce their emissions beyond specified regulatory requirements. To date, programs involving "emissions trading" have not been a critical part of state plans aimed at achieving attainment.[59]

Placing the primary responsibility for the development of plans on the states and local regions represents a sensible approach to environmental regulation. Within this context, it is important to develop a set of workable plans that will help meet the objectives mandated by Congress. Unfortunately, these plans are difficult to design and implement in some nonattainment areas. Most low cost technology-based standards have been applied. What remains are a series of high cost options. States, locales, and even the EPA are finding it increasingly difficult to induce industry to use options that are quite expensive and offer relatively modest gains in moving towards the standard.

The command-and-control approach is unlikely to make significant progress towards meeting the ozone standard in severe nonattainment areas. Thus, it is worthwhile considering new innovative approaches. One possibility is the use of market-based

---

[59] See Hahn and Hester (1986) and Liroff (1986). While these authors note that the overall use of trading programs has been limited, they also argue that the programs have resulted in significant cost savings to participants in the programs. The effect that these programs have had on overall environmental quality appears to have been small.

approaches similar to those promulgated under EPA's emissions trading programs (Hahn and Hester, 1986). The idea behind these approaches is to specify an overall ceiling for allowable emissions, but to allow firms a great deal of flexibility in choosing technologies to reduce emissions. In some instances it may be possible to design systems that have the potential to achieve cost savings and improve environmental quality in a timely manner. Whether these benefits can be achieved in practice depends on the nature of the systems that are put into place.

5.3. Designing and Evaluating Alternatives

The basic problem is to develop a framework that would enable nonattainment areas to meet the ambient standards, or substantially reduce their ozone concentrations in a timely manner. The subsequent analysis will take the goal of meeting the existing ambient ozone standard as a "given"; not because I believe this standard is justified, but because this is the way that Congress and EPA have chosen to frame the problem.

A major factor affecting the speed with which attainment can be reached is the cost of pollution control.[60] The question is whether it is possible to substantially reduce the control costs while maintaining or enhancing environmental quality. Here, I will argue that it is, indeed, possible to substantially reduce costs by placing appropriate incentives on firms to search for less expensive ways of reducing VOCs and nitrogen oxide emissions (NO_x), the two primary chemical species which contribute to the formation of ozone. The key idea is to design a regulatory system that promotes greater *flexibility* for individual firms while still making significant strides towards attainment of the ozone standard.

Since the emissions trading program is already in place, it is important to consider whether this program, will, in itself, be sufficient to promote attainment. The answer to this question is, unfortunately, no. At present, states have the option of whether they want to use some form of emissions trading as part of their attainment strategies. While elements of emissions trading are used in a large number of areas, the current program appears to have

---

[60] There are, of course, other factors and these will be considered shortly.

had little effect on environmental quality. This is, in part, because emissions trading represents a relatively small element of the existing regulatory approach. It is also because emissions trading is not explicitly designed to meet ambient standards, though parts of the program are aimed at reducing emissions. With suitable modifications, however, it would be possible to fashion an emissions trading program that would help address the ozone nonattainment problem.

Within the context of existing requirements for State Implementation Plans, there are two key criteria that need to be met for a program to be approved by EPA. The first is that the plan pinpoint identifiable emissions reductions from specific sources. The second is that it demonstrate that the projected emissions reductions will allow the area to meet the ambient standard for ozone. Since many states have plans that were approved by EPA, yet are still not in attainment, it is worth examining how this situation arose. As several authors have noted, there are many problems with the process (Hahn and Hester, 1986; and Liroff, 1986). There is a great deal of uncertainty in emissions inventories, and in the relationship between emissions and ambient air quality. It is also difficult to project economic growth and the effectiveness of different control options. Moreover, there are incentives for states to be overly optimistic in their projections. Thus, control options that are promised in the plan are not always implemented. Any alternatives that will be implemented should take these design requirements and uncertainties into account.

In devising any practical alternative, it is important to specify how decisions and responsibilities should be divided among state and federal regulators, and the Congress. For purposes of analysis, it will be assumed that Congress will specify a time frame in which the 0.12 ppm standard must be met. This time frame could vary across regions depending on the difficulties that are likely to be encountered in meeting attainment. One important consideration in setting a deadline or series of deadlines is the likelihood that regions will comply with the deadline. If deadlines are going to serve as more than symbolic gestures, it would be helpful if considerations of political and technical feasibility entered into the development of new legislation. Moreover, realistic mechanisms for enforcement will need to be added.

While Congress is in a good position to decide on general policy objectives, micromanagement is better left to federal and state agencies. One general problem that the states and EPA will need to grapple with is the definition of required emissions reductions of NO_x and VOCs. This exercise will need to be based on environmental modeling that predicts how ozone concentrations vary with different emission profiles. The general approach to this problem would be similar to the current approach taken with State Implementation Plans involving ozone control.

The problem of defining broad guidelines and general emissions targets applies to any program aimed at widespread environmental control. For specified applications involving market-based mechanisms, there are several other steps that need to be taken. At a general level, the rules for trading need to be specified. A good rule of thumb to follow is to encourage trading when it reduces costs, but does not compromise environmental quality. A key issue involving ozone attainment is whether trading should be allowed across different types of emissions. While EPA probably has the authority to allow interpollutant trading for ozone attainment, an endorsement from Congress would be helpful in expediting the development of this policy. Moreover, Congress may want to consider developing a framework that explicitly acknowledges interdependencies among pollutant problems. For example, NO_x emissions represent an important part of the acid rain problem as well as the ozone problem. By taking advantage of certain chemical interconnections, it will be possible to reduce the costs of achieving environmental goals.

Some of the most difficult political and administrative problems arise in designing the detailed applications of market-based alternatives. The options considered below will examine three different questions. First, how are emission rights defined and allocated? This is a critical question for determining who bears the costs of emission control, and how reductions are going to be achieved over time. Second, how are specific trading rules defined? This definition is critical in defining the scope for trading and the potential for cost savings and environmental quality improvements. Third, what type of administrative changes will be needed under the new system? This issue is important for assessing the practical feasibility of a proposal.

Three options are considered here. They all represent significant departures from the current policy. However, they differ in the degree to which they supplant the existing command-and-control approach. Each plan involves the reductions of VOC emissions and/or NO_X emissions. Trading across pollutant categories would be allowed under all plans. As noted above, previous control strategies have been defined primarily in terms of VOC reductions; thus, allowing trading between NO_X and VOC emissions would represent an important change in existing policy.

The first option consists of organizing a full-scale market in property rights for emission credits. To set up the market requires establishing a baseline for defining emission rights. States and local areas would be charged with defining the relevant baseline subject to broad guidelines set forth by EPA. Existing emissions inventories could be used to establish and revise this baseline. One alternative for establishing the baseline is to issue emissions rights for a specified period on the basis of actual emissions. This is similar to the approach that EPA has advocated in its final emission trading policy statement (51 *Fed. Reg.* 43814–43860). This allocation scheme is based on the grandfathering of emission rights. It takes the existing distribution of emissions as a legitimate basis for defining tradable rights. Firms that have already spent large sums of money on cleaning up pollution abatement may object to this system on the grounds that it rewards firms that have not shouldered their fair share of the burden. If these objections are strenuous, states may choose to establish other baselines that reflect local concerns. For example, some states might retain some rights for the public and sell them or auction them off to raise money for promoting environmental quality. Environmentalists could be expected to favor options that force industry to pay for polluting activities.

Once a realistic baseline is established for emission rights in both VOCs and NO_X emissions, a plan needs to be developed that yields the necessary emissions reductions. This can be accomplished relatively easily by reallocating emissions rights at specified intervals. For example, suppose that in the first year of the program, 100 rights were issued that allowed 100 tons of VOCs to be emitted. In year two, the objective is to cut these emissions in half. This could be done by redefining the value of existing rights, so that each right corresponds to 0.5 tons as opposed to 1 ton. In order to provide industry with some certainty, states would want to define the system for reducing rights with great care. The objective should be to provide firms with certainty while building enough flexibility in the system to adapt to changing environmental conditions (Hahn and Noll, 1982a).

It is important to recognize that the rules set up for reducing emissions can have an important effect on how individual firms respond to the market-based system. If firms that reduce emissions by more than the average are unduly penalized, this will make them think twice about exploring new technologies that could lead to any further emissions reductions. A simple way to avoid this problem is to dilute the value of existing emissions rights by some fixed fraction for all firms, independent of their previous performance.

A problem will also arise in situations where firms are emitting at levels that exceed those allowed by their emission rights. This problem is precisely analogous to the problem where an operating permit is violated. The regulatory agency will need to develop ways of dealing with such problems. A logical place to start is to refer these problems to the enforcement staff. It remains an open question as to whether the enforcement staff will need to be increased to accommodate a market-based regulatory initiative.

The states and local areas would develop plans subject to EPA approval. EPA should produce broad guidelines identifying the kinds of trades that are acceptable. For example, EPA could produce guidelines on trades involving mobile and stationary sources, and trades that involve more than one pollutant. In the case of interpollutant trading involving VOCs and NO_X, EPA should suggest appropriate modeling requirements for states. Decisions on specific trading rules should be left to the states. Like Congress, EPA should try to avoid micromanagement.

Note that this proposal would not distinguish between new and existing sources. New firms would not be required to meet any specific standard, but they would be required to own emission credits corresponding to their existing emissions. This is a dramatic departure from existing law, which imposes more stringent restrictions on new sources in the hopes that this will eventually result in better environmental quality.[61] Environmentalists have been strong supporters of legislation that imposes stringent requirements on new sources. They are unlikely to support this system unless they can be persuaded that it offers the potential for substantial improvements in environmental quality relative to the status quo.

Moving to a full-blown market approach would require some important changes in the way regulatory agencies do business. Engineering staff at state and local levels in charge of writing standards would now be asked to evaluate the validity of trades in terms of their environmental benefits. A record keeping system would need to be installed to keep track of changing ownership of emission rights. This system would rely primarily on the self-reporting of firms along with occasional audits.

One important issue relates to the definition of valid trades. To promote cost savings, trading should be defined as broadly as possible. For example, electric utilities that show demonstrable gains in environmental quality as a result of conservation efforts should receive the same treatment as utilities that choose to install hardware, such as scrubbers. The precise details of trading rules should depend on key parameters related to the transport and formation of ozone. States and local areas should be given a broad mandate so they can tailor trading programs to their individual needs and capabilities.

At the federal level, EPA would have to increase its oversight function. It would want to focus on the validity of baseline estimates as well as evidence that the state was heading towards its attainment goals. EPA could accomplish this oversight through occasional auditing of individual trades; asking for updates on aggregate emissions of NO_x and VOCs; and using data from existing monitoring networks.

This proposal would face many hurdles in implementation. However, none of these hurdles is insurmountable. The problem of establishing an appropriate baseline is already addressed under current State Implementation Plans. There is, admittedly, a great

---

[61] In reality, the effect of the new source legislation has been to induce many firms to retain outdated plant and equipment for a longer time period than they otherwise would have.

deal of uncertainty in emission baselines and inventories.[62] This does not imply, however, that trading cannot work effectively any more than it implies that command-and-control approaches cannot work effectively. What it does suggest is that the design of a trading system needs to take the quality of the existing data into account.[63]

There are several ways that safeguards can be built into the system. At an aggregate level, if specific interim targets related to air quality were not met, the plan could call for greater across-the-board reductions in emissions rights. At the level of individual trades, EPA could expand its role as an auditor, in much the same way that the Internal Revenue Service does. Instead of trying to monitor each individual trade in detail, for example, it could monitor selected trades. Over the long run, there is no substitute for getting a better picture of emissions profiles for individual firms and consumers. However, in the past, Congress and the EPA have demonstrated a marked unwillingness to address this problem.

Another potential objection to this plan is that it runs counter to the intent and legislative requirements of the Clean Air Act. Under current law, there is a distinction between new and existing sources. Sources that were in existence when emissions were first inventoried in the mid-1970's are called existing sources; sources that have been built since that time are called new sources. The distinction between new and existing sources is important because new sources generally must comply with more stringent technology-based regulations. Existing sources have the option of meeting emissions standards through conventional approaches or by purchasing emissions credits. If Congress is unwilling to lift the restrictions on new sources, then this proposal may be infeasible.

As an alternative, consider a second option that requires all new sources to meet the more stringent control requirements, but is otherwise identical to the first proposal. Relative to the first option, this proposal could be expected to reduce the amount of new sources locating in nonattainment areas since it would be more expensive to locate there. Additionally, the overall level of cost savings would decline, since new sources would have less flexibility

---

[62] The recent EPA emissions trading policy statement attempts to address these issues (51 *Fed. Reg.* 43814-43860).

[63] For a somewhat more pessimistic assessment of the potential for trading, see Meiburg (1987).

in how control requirements were met. The major advantage of this proposal in comparison to the first proposal is political. Part of the reason for the existing new source legislation is to protect eastern coal mining jobs. This proposal would retain the regulations that protect these jobs. At the same time it would add much greater flexibility into the existing system by changing the structure of the regulatory system to promote trading. The structural changes would be precisely the same as those discussed in the previous option.

Both the first and second options represent radical departures from the *status quo*. Moreover, they tend to accentuate the role of markets. If there is anything that recent lessons from the application of market-based approaches has taught us, it is that full-blown markets tend to be the exception rather than the rule. This is especially true when there is a great deal of controversy over the appropriate distribution of property rights, and it is difficult to monitor and enforce standards. For precisely these reasons, it is unlikely that these two approaches will be implemented. Thus, it is worthwhile considering policies that are more in line with the existing regulatory approach.

The third option consists of building directly on EPAs current strategy for dealing with the ozone problem. EPA currently identifies control strategies that must be used for both new stationary sources and mobile sources. For example, recently there has been a debate over whether to install devices in gas tanks which reduce emissions from refueling. The control strategy is referred to as "onboard" (U.S. EPA, 1984). The automobile companies are understandably concerned about the cost and effectiveness of this option. Yet, under the current system, their options are quite limited if EPA decides to implement this approach. They must comply with the technology-based standard or be in violation of the law.

An alternative approach is to continue to allow EPA to impose technology-based standards, but to allow companies to meet the standards by making equivalent or greater emissions reductions through other means. This concept is very similar to the EPA's existing "bubble policy". The bubble policy allows a firm to sum the emission limits from individual sources of a pollutant in a plant, and to adjust the levels of control applied to different sources so long as this aggregate limit is not exceeded. The bubble policy, as currently implemented, applies to existing sources. The approach considered here is precisely analogous to the bubble concept, except that it applies to new regulations on both mobile and stationary sources.

Consider the example of the onboard system. Suppose the automobile companies found a new, less expensive, technology that significantly reduced emissions, such as a new catalytic converter or a new fuel. Then, they should be allowed to implement this technology instead of the federally mandated solution provided it meets or exceeds the environmental targets that the regulators had in mind when setting the standards. Even if the companies do not make the reductions themselves, they should still be allowed to receive credit for them, provided they can persuade other companies to make changes that result in verifiable emissions reductions. Thus, for example, the automobile companies might find it less costly to pay oil refiners to reduce emissions than to install the onboard system themselves.

There is no reason why this third option could not be extended to specific policies mandated by states to meet the ozone standard as well. In any instance where regulators promulgate a technology-based standard, firms could be given the opportunity to identify and implement an approach that either is less costly or achieves equivalent or greater improvements in environmental quality.

The third option would still require major emission reductions in some nontattainment areas. States and local areas would have the primary responsibility for regulations in these areas. However, EPA could augment these reductions with regulatory requirements and guidelines, such as onboard. Note, however, that EPA need not pursue a strategy that requires across-the-board emission reductions in all areas. Rather, it should continue to focus its efforts on non-attainment areas. Difficulties will inevitably arise in targeting mobile source reductions. However, these difficulties do not imply that the best strategy is necessarily to require all mobile sources to install stringent control equipment.

One of the attractive features of this approach is that it does not require major administrative changes. EPA would continue to regulate ozone within the existing framework. The only significant change would be the added flexibility given to firms in meeting new standards. Yet, this change is important because it could increase the political acceptability of the current approach by allowing firms the opportunity to save money by implementing less costly control technologies.

The proposals presented here are not without their problems. Moreover there are many details that would need to be worked out in implementing these ideas. Since design issues have been addressed in detail by several authors, the focus here will be on those issues that are likely to be of greatest concern for the particular problem at hand.[64]

One important issue is how to compare emissions reductions from different kinds of sources. In this area, there are likely to be important tradeoffs between administrative simplicity and the degree to which the regulatory approach is fine-tuned to address specific features of individual sources. It is clear that certain kinds of emissions reductions are subject to greater uncertainties than others. For example, stationary source reductions may be more easily estimated and verified than emissions reductions involving mobile sources. Even within the stationary source category, the ability to verify reductions will differ across sources. It would be cumbersome and unnecessary to define emission credits that differ in value for each abatement technology. Unless there are pressing reasons for considering reductions from one source as being less effective or less certain than similar reductions from another source, the two reductions should be weighted equally by the regulator. This is not to suggest that all reductions of a particular type of emissions be treated equally. Attention can and should be given to the costs of monitoring, and the distribution of uncertainty over claimed emission reductions. When agency monitoring costs are very high, or uncertainty is quite large, this might constitute grounds for either diluting the value of a particular credit, or alternatively, asking the source proposing the reduction to address these issues constructively.

A related issue is how to design an overall scheme for achieving reductions in ozone (Hahn, McRae and Milford, 1985). As noted earlier, both VOCs and NO_X emissions contribute to the formation of ozone, and they do so in a highly nonlinear manner. It is by no

---

[64] For a discussion of general design issues, see Hahn and Noll (1982a, 1982b) and Tietenberg (1985). For an evaluation and discussion of concrete policy proposals related to emissions trading, see Hahn and Hester (1986) and Liroff (1986).

means obvious that the best approach for addressing the ozone problem in specific areas is to continue to reduce VOC emissions. States need to reevaluate this strategy. All of the proposals suggested above can be used to help meet changes in the required levels of NO_x or VOX emissions.[65]

The options examined here are not foolproof. Thus, it is reasonable to ask what happens if a mistake is made. The answer to this question depends on the source of the error and the size of the problem it creates. Probably, the most serious problem relates to missing the overall target for emissions reductions. As noted earlier, this can be accommodated in the first two options by further tightening of the overall number of available emission credits. Under the third option, EPA would have to continue imposing more stringent measures, and then allow firms to search for alternative ways of meeting the prescribed emissions limits.

All three of these proposals are designed to set up a system of incentives that directly involves industry in a constant search for more productive ways of cleaning up environmental problems. The specific details of the proposals are meant to be suggestive. For example the appropriate method for allocation of permits would be determined by the political process. Grandfathering is suggested here only because the existing distribution of property rights frequently has an important effect on the design of new regulatory systems (Rolph, 1983; Welch, 1983). The only part of the design that is critical is allowing firms greater flexibility in meeting prescribed environmental targets.

Given the promise of these proposals, it is reasonable to consider why they have not been implemented. One problem is legal. The first option may not be allowed because it would no longer require new sources in nonattainment areas to meet specific emission targets. However, the status of new source trading is changing. EPA recently approved a bubble that would allow trades between two generators that are subject to the same new source regulation (50 *Fed. Reg.* 3688–3695). The next logical step would be to consider bubbles between new and existing sources.

---

[65] For a proposal that would allow regulators to trade off among NO_x and or VOC emissions see Hahn (1987a).

The other two options appear to be well within the realm of what EPA and states can do within the current bounds of the Clean Air Act. Thus, the reasons for not using these options must lie elsewhere. To understand the reasons for not exploring these options, it is useful to look at the past attitudes of key actors in the decision making process. At a federal level, neither Congress nor EPA has ever been terribly disposed towards moving away from command-and-control approaches. While some states have employed market-based approaches, state and local agencies have not generally promoted this option, except in a few circumstances. The past attitudes of industry towards trading is characterized by a healthy skepticism. There has been concern that these approaches might not confer significant benefits, and might introduce considerably uncertainty. Environmentalists have vigorously opposed most attempts at initiating or expanding the scope for trading, arguing that trading is very difficult to implement without resulting in adverse environmental consequences (Liroff, 1986; National Clean Air Coalition, 1985).

The most vigorous support for previous trading initiatives has come from a surprisingly small set of individuals. In the case of emissions trading, the primary support for programs came from a small group of scholars and reform-minded bureaucrats (Hahn and Hester, 1986; Meidinger, 1985).

Is there any reason to think the views of various groups will change? I believe there is. As it becomes increasingly apparent that the command-and-control options are not buying very much, both industry and environmentalists may become more receptive to different alternatives. Legislators will be left with three basic choices. The first is to "redefine" the standard in areas where it can't be met. The second is to continue along the current path, which is unlikely to buy very much, but gives an outward appearance that something constructive is happening. The third is to explore new approaches along the lines suggested here.

If new approaches are to be promoted, in all likelihood, Congress will have to provide greater incentives for both EPA and the states to explore these ideas. The key signal that Congress will need to send out is that flexibility is going to be encouraged in meeting the goals of the Clean Air Act. For example Congress could require, at a minimum, that EPA allow firms to use alternative approaches for meeting their proposed standards. Moreover, it could suggest that alternative approaches be viewed in the same context as the current approach to regulation. There is currently a striking asymmetry between emission credits that are traded and those that are not traded in terms of how they are treated (Hahn and Hester, 1987b). Emissions trading effectively establishes two classes of emission rights: those that are traded and those that are not. State regulators have tended to treat traded rights in a way that affords them an inferior status. This can be seen in the treatment of banked credits in selected banking programs. These credits are sometimes subject to a "discount" which is a *de facto* partial confiscation. The inferior status of traded rights also can be seen in the close scrutiny regulators give to the creation of emission credits. Eliminating this asymmetry in the treatment of credits would help induce firms to search more vigorously for cost saving environmental improvements. All three options presented above could be tailored so that traded rights could stand on an equal footing with untraded rights.

Related to this idea, Congress could require that EPA spell out the type of tradeoffs it will allow among different types of technologies. For example, how much credit will given for using cleaner fuels in automobiles, such as methanol, or different blends of gasoline? Currently, firms do not have very much incentive to explore such options. This incentive could be increased by spelling out the rules that apply in counting the environmental benefits from different approaches.

Spelling out these tradeoffs will not be easy in some cases. Moreover, the task is complicated by the fact that many technologies do not offer unambiguous improvements for all pollutant categories. These technologies could be precluded from being adopted, as is typically done now. However, this strategy may not make sense either from the perspective of increasing overall environmental quality or reducing costs.

Congress should also consider encouraging EPA to specify conditions under which trading could take place between NO_x and VOC reductions. As noted earlier, this is a tricky problem, in that the specific schemes adopted will need to be tailored to air quality conditions in specific regions. EPA would need to provide guidance to the states about what sort of trade-offs are permissible.[66]

An important strategic issue in addressing the ozone nonattainment problem is the extent to which controls are targeted for nonattainment areas. In the past, Congress and the EPA have been reluctant to promote standards for mobile sources that differ by regions. However, as noted above, targeted strategies may be better than across-the-board standards if the problem is viewed as regional. Thus, even if EPA's onboard strategy were a good idea, it need not be required for all cars if the sole objective is to reach attainment in areas that currently exceed the standard. As noted above, a critical issue in designing the mobile source component of an ozone control strategy is to ensure that reductions in emissions are verifiable. This problem can be handled in several ways. Credits for emission reductions could be based on estimated emissions associated with different types of mobile source categories. For example, sales of vehicles to owners of dedicated fleets, such as taxicabs, could result in higher levels of emission credits per vehicle than sales to the public at large.

Even if Congress came out unambiguously in favor of market-based alternatives that promote meeting the standard, it is possible that very little would actually be done to enhance environmental quality. There are two reasons why this outcome might arise. One is that the market-based programs could be designed poorly. A second, and probably more likely one, is that the current system of rewards and sanctions will serve as a major impediment to getting states to move in a timely manner. The rewards for being in attainment are not terribly high; nor are the credible sanctions available to EPA for inducing progress in nonattainment areas. The current set of draconian sanctions related to highway funds, sewage treatment funds, and construction bans needs to be replaced by measures that are less blunt. Administrators have been very reluctant to impose these sanctions because they have potentially dire economic and political consequences. In addition to exploring

---

[66] California has already begun to use interpollutant trades on a limited basis for a variety of pollutants (Menebroker, 1987). Examples include trades of NO_x for VOCs, and trades of particulate matter for NO_x, VOCs, and sulfur oxides.

the modification of sanctions, Congress may want to reconsider the amount of discretion the Administrator has in imposing particular sanctions. If sanctions and rewards are viewed as credible, then states are more likely to take them seriously.

If EPA were given a signal to promote flexibility, there are several steps it might take. Consistent with the suggestions above regarding legislative changes, it could provide greater guidance on tradeoffs so firms had a better sense of the payoffs that would result from adopting innovative technologies. Another way of increasing trading activity is to decrease the level of federal oversight on particular trades. EPA's "generic bubble" policy, which allows states to approve bubbles without direct federal oversight, represents a step in this direction. The agency could also establish a series of rules regarding trading between new and existing sources in nonattainment areas, it this were not spelled out in the enabling legislation.

This section has argued that the current Clean Air Act is not sufficient to promote widespread use of markets as a vehicle for expediting attainment. Changes will be required in the Clean Air Act and in the way EPA interacts with the states if market-based approaches are going to play a significant role in addressing the ozone nonattainment problem. While the second and third options outlined above could probably be implemented without any legislative changes, it is unlikely that they will be adopted in the current political environment.

5.4. Conclusions and Areas for Future Research

This section has illustrated how to apply some of the basic ideas developed in the preceding sections to a concrete policy problem. Design of systems that makes sense and fall within the realm of feasibility requires an understanding of political institutions, political forces, economics and science.

Economists are quick to point out that "there is no free lunch." In some ways, though, there is a free lunch out there. It is possible to decrease costs and increase environmental quality through the judicious design and implementation of new regulatory approaches. Of course, what is possible and what will actually happen are two very different things. This explains, in part, why attempts to initiate market-based approaches have met with such strong resistance. Many groups simply don't believe that greater reliance on the market approaches will work. Some groups also question whether these approaches represent an appropriate response to environmental problems.

Appropriate or not, I have tried to argue that more flexible approaches may be one of the few ways that can be used to help achieve the ozone standard. These approaches are not without their pitfalls, but then neither is the existing system.

This section has focused exclusively on market-based approaches. However, there is no reason, in principle, why other incentive-based options, such as emissions fees, could not be considered as well. For example, firms could be permitted to pay a fee to the state in lieu of obtaining emission reduction credits directly. This fee could then be used to enhance environmental quality. The firm paying the fee would be given emission credits based on the expected environmental improvements that would result from using revenues from the fee. In addition to enhancing environmental quality directly, the fee could also be used to improve emissions inventories and monitoring capabilities. This is an area that deserves further exploration.

Given current political interests, it is unlikely that full-blown markets will be used to address the ozone nonattainment problem. It is possible, but unlikely that a more incremental approach, such as the limiting trading described in the third option, will be used. As it becomes more apparent that the political costs of reaching attainment are quite high, Congress will be left with some difficult choices, including effectively rescinding the current standard for selected areas. If Congress is serious about wanting to use market approaches, legislation will be needed to help promote their use.

6. WHITHER ENVIRONMENTAL REGULATION?

Environmental expenditures now are in the billions of dollars in several developed countries. Moreover, they can be expected to increase in the future. This reflects growing public concern about environmental issues in developed countries. This concern can be expected to be mirrored in developing countries, especially those that are enjoying rapid rates of development. The public increasingly expects the government to play a more central role in managing environmental problems.

There are several options that are available to regulators and politicians for addressing environmental concerns. This book has focused on "action-oriented" options that attempt to change the overall level of emissions and resulting environmental quality. The instruments considered here provide markedly different incentives for firms. Standards, the dominant instrument on the environmental regulatory scene, appear not to do well in terms of keeping expenditures down. However, this statement is based more on a theoretical assessment than a practical evaluation of their relative performance against other instruments. The problem is that it is difficult to obtain data that would allow the direct comparison of different types of instruments. Moreover, as noted in Section 3, almost all regulatory systems tend to be mixed in nature, thus making it difficult to isolate the effects of a particular instrument.

In contrast to standards, incentive-based instruments such as emissions fees and marketable permits are thought to have desirable efficiency characteristics. However, a practical evaluation of actual applications revealed a somewhat different story. Only in rare instances have existing emission fees been shown to have marked effects on firm behavior. The fees primarily serve as a mechanism for raising revenues, which are then earmarked for environmental quality improvements. Market-based approaches, on the other hand, have not had a dramatic impact on environmental quality, but have resulted in sizable cost savings.

Both sets of instruments exhibit wide variation across different applications. Much of the variation in the performance of these instruments is consistent with economic theory. The higher the fee, and the more directly it is linked to current performance, the more likely it will have an effect on emissions. Similarly, the lower the transactions costs in the permit market, the more likely it is that trading will occur, and cost savings will result.

In addition to explaining the relative performance from an economic perspective, this book attempted to examine why different instruments are selected. The analysis relied heavily on an interest group paradigm in which environmentalists were pitted against industry. This analysis revealed that, only in rare instances, are markets resembling those in textbooks likely to be implemented in practice in the near future. The application that comes closest to a textbook description is the case of lead trading, which had several characteristics that set it apart from other programs such as emissions trading.

The interest group paradigm was not only useful in explaining narrow issues in instrument choice, but also useful in providing insights into broader patterns of regulation. For example, the relationship between breadth and stringency of regulation, and the use of targets as symbols, both can be explained using this framework. Indeed, this paradigm can be useful in understanding what is likely to be feasible in the current political environment. In this sense the theory is not only useful for explaining past events, but it also has predictive power. For example, in the application to the ozone nonattainment problem, it was possible to assess the likelihood that different types of market-based approaches would be adopted. In general, chances for adoption are quite low. However, the possibility of incremental approaches that make greater use of markets cannot be ruled out.

The standards-based approach to environmental regulation is here to stay for the foreseeable future. Unfortunately, not that much is known about the behavior of standards in practice. For example, precisely what effects do they have on innovation? How do different types of standards, such as performance and technology-based standards, differ in their effects? It is possible to answer some of these questions in theory, but given the discrepancy between theory and practice, it behooves us to take a closer look at some of these issues before assuming that we really understand what is going on.

The same set of comments regarding standards also pertain to subsidies and grants. Section 2 noted the key role that grants and subsidies play in EPA's budget. These instruments represent an important and poorly understood aspect of environmental regulation. Again, there is some theory that can guide us, but precious little data on comparative performance of different types of subsidies and grants.

While standards and subsidies will not disappear from the scene, the role for incentive-based options can be expected to increase. If the analysis suggested here is correct, these tools have had a salutary impact on environmental problems. They have tended either to reduce costs and/or increase environmental quality. Although wider acceptance of incentive-based instruments is certainly not guaranteed, it is likely to be forthcoming unless specific applications of these instruments are associated with events that have particularly adverse environmental consequences.

The U.S. EPA is exploring different ways of expanding the use of market mechanisms for emission trading. The trade between two utilities subject to new source regulation is but one example. EPA is currently examining the possibility of expanding the scope for trading between new and old sources. It also is considering possible applications to the problem of acid rain, where there is the potential to save a great deal of money while maintaining or improving environmental quality. In that case, one of the major problems is how to preserve jobs for coal miners in selected areas of the country (Ackerman and Hassler, 1981).

There is no reason, in principle, why incentive-based mechanisms could not be applied to more general categories of risks. For example, the risk reductions associated with hazardous waste sites could be traded off against risk reductions associated with abatement strategies aimed at reducing ozone. Economists have argued that such trading could lead to dramatic efficiency gains (Hahn, 1986; Portney, 1988). Merits of this argument notwithstanding, politicians and regulators are likely to be less sanguine about making such trade-offs explicit, even if they could agree on an acceptable definition of risk. The public has only reluctantly accepted incentive-based mechanisms in a limited framework. Expanding that framework will not be easy.

One of the issues only briefly touched on in this book that deserves further exploration is the linkage between monitoring and enforcement and instrument choice. Just like marketable permits and emission fees, the system of monitoring and enforcement can have an important impact on program performance. Moreover, the ability to monitor and enforce certain instruments has had an impact on the views of different instrument groups. For example, environmentalists have frequently criticized emissions trading on the grounds that it represents another regulatory loophole for industry. Part of this argument stems from difficulties that are perceived to be attached to demonstrating that trades have a salutary effect on environmental quality.

The evaluation of instruments and the application of this knowledge to a specific problem in air pollution now permits an answer to one of the central questions posed in this book: Is it possible to get more for less? The answer to this question is a conditional yes. The performance of both emissions trading and lead trading indicates that large cost savings can result without having a significant effect on environmental quality. This suggests that it is possible to achieve comparable levels of environmental quality with less money. However, is it possible to get better environmental quality with fewer resources? This was a key issue that was examined in the preceding section on ozone nonattainment. My belief is that it is certainly possible to design such schemes. Whether they will be politically feasible is another question. However, just because existing market-based systems do not appear to have had a major impact on environmental quality does not imply that these systems can't or won't have a salutary impact in the future. The application at Dillon Reservoir could be expected to have just such an effect if trading ever gets off the ground there.

The design of new regulatory approaches for environmental problems is a difficult task. The analysis of the ozone nonattainment problem underscored the importance of understanding the political institutions. It also demonstrated the importance of having a good grasp of the science. Armed with a specific knowledge of the problem at hand, it will then be possible to apply some of the general insights offered in this book to help design systems that are simple, workable, and feasible.

Appendix

This appendix contains formal proofs of the propositions in Section 4. Because the proofs of the various propositions are quite similar, some proofs of later propositions are outlined.

PROPOSITION 1: *An increase in industry influence will increase the market orientation of the instrument and reduce the level of environmental quality which is selected.*

PROOF: An increase in industry influence is represented by an increase in a. It suffices to show that $\partial M/\partial a > 0$ and $\partial Q/\partial a < 0$. Total differentiation of the first order conditions resulting from (4.1) yields:

$$\begin{bmatrix} aI_{11} + (1-a)E_{11} & aI_{12} + (1-a)E_{12} \\ aI_{12} + (1-a)E_{12} & aI_{11} + (1-a)E_{11} \end{bmatrix} \begin{bmatrix} dM \\ dQ \end{bmatrix} = \begin{bmatrix} (E_1 - I_1)da \\ (E_2 - I_2)da \end{bmatrix}$$

This problem can be solved by inverting the matrix. The assumptions on the preferences of environmentalists and industry insure that the matrix of second-order derivatives is negative definite. Inversion yields the following sign pattern:

$$\begin{bmatrix} dM \\ dQ \end{bmatrix} = \begin{bmatrix} - & +/0 \\ +/0 & - \end{bmatrix} \begin{bmatrix} -da \\ +da \end{bmatrix}.$$

This sign pattern implies $\partial M/\partial a > 0$ and $\partial Q/\partial a < 0$.

PROPOSITION 2: *If preferences are well-behaved, an increase in industry influence will result either in a decrease in environmental quality and/or an increase in the market orientation of the instrument.*

PROOF: The proof relies on the fact that the function being maximized is negative definite. Note that the effect of a change in a on industry utility is given by the expression:

$$\frac{\partial I}{\partial a} = I_1 \frac{\partial M}{\partial a} + I_2 \frac{\partial Q}{\partial a}.$$

Since $I_1 > 0$ and $I_2 < 0$, it suffices to show that $\partial I/\partial a > 0$.

Define "det A" to be the determinant of the matrix of second order partials. Solving explicitly for the effects of a change in a yields:

$$\frac{\partial M}{\partial a} = (1/\det A)[(aI_{22} + (1-a)E_{22})(E_1 - I_1)$$
$$- (aI_{12} + (1-a)E_{12})(E_2 - I_2)],$$

and

$$\frac{\partial Q}{\partial a} = (1/\det A)[-(aI_{12} + (1-a)E_{12})(E_1 - I_1)$$
$$+ (aI_{11} + (1-a)E_{11})(E_2 - I_2)].$$

Multiplying $\partial M/\partial a$ by I_1 and $\partial Q/\partial a$ by I_2, and adding gives the following expression for $\partial I/\partial a$:

$$(a/\det A)\{I_2(E_2 - I_2)I_{11} - [(E_1 - I_1)I_2 \\ + (E_2 - I_2)I_1]I_{12} + I_1(E_1 - I_1)I_{22}\} \quad (4.5)$$

Since I is strictly concave and twice differentiable, the quadratic form associated with the Hessian of I is negative definite. This implies that the associated quadratic form is negative. Through suitable manipulation, (4.5) can be related to a quadratic form. The first order conditions associated with (4.1) imply:

$$E_j = [a/(1-a)]I_j \quad \text{for} \quad j = 1, 2.$$

Substitution into (4.5) yields:

$$[a/((a-1)\det A)]\{I_2^2 I_{11} - 2I_1 I_2 I_{12} + I_1^2 I_{22}\}$$

after factoring $(1/(a-1))$. The first expression, $[a/((a-1)\det A)]$, is less than 0 since $a \in (0, 1)$ and $\det A > 0$. The second expression is a quadratic form. To see this define the vector $(h_1, h_2) = (I_2, -I_1)$. Then the bracketed expression takes the form $h_1^2 I_{11} + 2h_1 h_2 I_{12} + h_2^2 I_{22}$. Since $(h_1, h_2) \neq 0$, and the quadratic from is negative definite, this implies that the expression in brackets is less than 0. Multiplying the two negative expressions together yields the result that $\partial I/\partial a > 0$.

PROPOSITION 3: *As uncertainty associated with any level of market orientation declines, it becomes more likely that a market-based approach will be chosen.*

PROOF: The problem is to show that the standard-based solution, where $M = 0$, is less likely to be selected. The proof of this proposition follows almost immediately. Consider the maximization problems of the form given by (4.2). Now substitute two new concave functions Φ' and Ψ' which have the property $\Phi' \geq \Phi$ and $\Psi' \geq \Psi$. Because the substitution of these functions will not decrease the value of the maximand for any values of M and Q, this implies that an interior solution to this problem will result in an increase or no change in the value of the objective function. If the objective function increases, then it is more likely to dominate the *status quo*.

PROPOSITION 4: *If preferences are independent, then the optimal level of M will decline upon introducing uncertainty.*

PROOF: The proof of this proposition requires that the first order conditions for (4.1) be compared with the first order conditions for (4.2). The first order conditions for (4.2) are:

$$a(I_1 + \Phi_1) + (1 - a)(E_1 + \Psi_1) = 0.$$

and

$$aI_2 + (1 - a)E_2 = 0.$$

Suppose (M^*, Q^*) satisfies the first order conditions for (4.1). This will not satisfy both first order conditions for (4.2) because the uncertainty terms enter into the first of these conditions. Since $\Phi_1 < 0$ and $\Psi_1 < 0$, this implies that when this condition is evaluated at (M^*, Q^*), it will be less than 0. Decreasing M will increase the left hand side of this first order condition. This follows from the assumption on the second order derivatives. Suppose (M^{**}, Q^*) satisfies the first of these first order conditions, where $M^{**} < M^*$. Note that this solution also satisfies the second first order condition for (4.2) under the assumption that $I_{21} = E_{21} = 0$. This completes the proof. Note that under the assumptions here, with $\Phi(0) = \Psi(0) = 0$, it follows that $M^{**} = 0$. However, if Φ and Ψ were positive, M^{**} could be positive.

PROPOSITION 5: *An increase in the relative influence of industry will result in a decrease in fees if preferences are independent. An increase in the relative influence of industry will result in a decrease in fees and no change or an increase in earmarking if the cross partials are non-negative and the marginal utility of earmarking for environmentalists does not exceed the marginal utility of earmarking for industry.*

If preferences are well-behaved, an increase in industry influence will result in a decrease in fees and/or an increase in earmarking.

The first part of the proposition can be derived by totally differentiating the first order conditions. The results, not shown here, are the same as for Proposition 1, except that M and Q are replaced by F and U. Assuming the cross partials for industry and environmentalists are 0 yields the following sign pattern:

$$\begin{bmatrix} dF \\ dU \end{bmatrix} = \begin{bmatrix} - & 0 \\ 0 & - \end{bmatrix} \begin{bmatrix} + da \\ ?da \end{bmatrix}.$$

This implies $\partial F/\partial a < 0$.

Assuming that the cross partials are non-negative and the marginal utility of earmarking for environmentalists does not exceed the marginal utility of earmarking for industry yields:

$$\begin{bmatrix} dF \\ dU \end{bmatrix} = \begin{bmatrix} - & +/0 \\ +/0 & - \end{bmatrix} \begin{bmatrix} + da \\ -/0da \end{bmatrix}.$$

This sign pattern implies $\partial F/\partial a < 0$ and $\partial U/\partial a \geq 0$.

The proof used to show that an increase in industry influence will result either in a decrease in fees and/or an increase in earmarking is precisely analogous to the proof used for Proposition 2 and will not be repeated here.

PROPOSITION 6:   *If preferences are independent, then an increase in industry influence decreases the actual level of environmental quality along with the target level of environmental quality. If preferences ae well-behaved, then an increase in industry influence will either decrease actual environmental quality and/or the target level of environmental quality.*

PROOF   Assuming the cross partials are 0 yields the following sign pattern:

$$\begin{bmatrix} dQ \\ dS \end{bmatrix} = \begin{bmatrix} - & 0 \\ 0 & - \end{bmatrix} \begin{bmatrix} + da \\ + da \end{bmatrix}.$$

This sign pattern implies $\partial Q/\partial a < 0$ and $\partial S/\partial a < 0$.

In the case in which the cross-partial derivatives cannot be signed, the proof is precisely analogous to that of Proposition 2.

COROLLARY 1: *If preferences are independent, then an increase in industry influence decreases the breadth and stringency of regulation. If preferences are well-behaved, then an increase in industry influence will either decrease breadth and/or the stringency of regulation.*

PROOF: The proof of Corollary 1 is identical to the proof of Proposition 6 since the same underlying structure is assumed.

References

Ackerman, B. and W. Hassler, (1981), *Clean Coal/Dirty Air,* Yale University Press, New Haven, Connecticut.

Allison, G. (1971), *The Essence of Decision,* Little, Brown and Co., Boston, Massachusetts.

Aranson, P., Gellhorn, E. and Robinson, G. (1982). "A Theory of Legislative Delegation,"*Cornell Law Review,* **68,** 1-67.

Arnold, R. (1979), *Congress and the Bureaucracy,* Yale University Press, New Haven, Connecticut.

Badaracco, Joseph (1985), *Loading and Dice: A Five-Country Study of Vinyl Chloride Regulation,* Harvard Business School Press, Boston, Massachusetts.

Bailey, E. (1986), "Deregulation: Causes and Consequences." *Science,* **234,** December 5, 1211-1216.

Barde, J. (1986). "Use of Economic Instruments for Environmental Protection: Discussion Paper," ENV/ECO/86.16, Organization for Economic Cooperation and Development, September 9, 27 pp.

Baumol, W. and Oates, W. (1975). *The Theory of Environmental Policy.* Prentice-Hall, Englewood Cliffs, N.J.

Becker, G. (1983). "A Theory of Competition Among Pressure Groups for Political Influence," *Quarterly Journal of Economics,* XCVII, 371-400.

Bohm, P. and Russell, C. (1985), "Comparative Analysis of Alternative Policy Instruments," In *Handbook of Natural Resource and Energy Economics.* Volume I, edited by A. Kneese and J. Sweeney, Elsevier Science Publishers, New York, New York, 395-461.

Boland, J. (1986), "Economic Instruments for Environmental Protection in the United States," ENV/ECO/86.14, Organization for Economic Cooperation and Development, September 11, 83 pp.

Borenstein, S. (1985), "Price Discrimination in Free-Entry Markets," *Rand Journal of Economics,* **16,** 380-397.

Bower, B. *et al.* (1981), *Incentives in Water Quality Management: France and the Ruhr Area.* Resources for the Future, Washington, D.C.

Bressers, J. (1983). "The Effectiveness of Dutch Water Quality Policy." Twente University of Technology, Netherlands, Mimeo. 31 pp.

Brickman, R., Jasanoff, S. and Ilgen, T. (1985), *Controlling Chemicals: The Politics of Regulation in Europe and the United States,* Cornell University Press, Ithaca, New York.

Brown, G., Jr. (1984a). "Economic Instruments: Alternatives or Supplements to Regulations?." *Environment and Economics,* Issue Paper, Environment Directorate OECD, June, 103-120.

Brown, G., Jr. (1984b). "Selected Economic Policies for Managing Hazardous Waste in Western Europe," mimeo, prepared for the Environmental Protection Agency, August, 36 pp.

Brown, G., Jr. and Bressers, J. (1986). "Evidence Supporting Effluent Charges," mimeo, September, 28 pp.

Brown, G., Jr. and Johnson, R. (1984), "Pollution Control by Effluent Charges: It Works in the Federal Republic of Germany, Why Not in the U.S.," *Natural Resources Journal*, **24**, 929–966.

Buchanan, J. and Tullock, G. (1975), "Polluters' Profits and Political Response: Direct Controls Versus Taxes," *American Economic Review*, **65**, 139–147.

Campos, J. (1987), "Toward a Theory of Instrument Choice in the Regulation of Markets," California Institute of Technology, Pasadena. California, Mimeo, January 26. 30 pp.

Coelho, P. (1976), "Polluters' Profits and Political Response: Direct Control Versus Taxes: Comment," *American Economic Review*, **66**, 976–978.

Cohen, M. and Rubin, P. (1985), "Private Enforcement of Public Policy," *Yale Journal on Regulation*, **3**, 167–193.

Dales, J. (1968), *Pollution. Property and Prices*, University Press. Toronto, Canada.

David, M. and Joeres, E. (1983), "Is a Viable Implementation of TDPs Transferable?," in E. Joeres and M. David. eds., *Buying a Better Environment: Cost-Effective Regulation Through Permit Trading*, University of Wisconsin Press, Madison. Wisconsin, 233–248.

Derthick, M. and Quirk, P. (1985). *The Politics of Deregulation*. The Brookings Institution, Washington, D.C.

Dewees, D. (1983), "Instrument Choice in Environmental Policy," *Economic Inquiry*, *XXI*, 53–71.

Dowlatabadi, H. and Hahn, R. (1986), "Approaches to Environmental Regulation," *Science*, **233**, 990–991, August 29.

Edelman, M. (1964). *The Symbolic Use of Politics*. University of Illinois Press. Champaign, Illinois.

Elmore, T. *et al.* (1984), "Trading Between Point and Nonpoint Sources: A Cost Effective Method for Improving Water Quality," paper presented at the 57th annual Conference/Exposition of the Water Pollution Control Federation, New Orleans, Louisiana, 20 pp.

Farber, K. and Rutledge, G. (1986), "Pollution Abatement and Control Expenditures," *Survey of Current Business*, 94–105, July.

Farber, K. and Rutledge, G. (1987), "Pollution Abatement and Control Expenditures, 1982–1985," *Survey of Current Business*, 21–26, May.

Fenno, R. (1973), *Congressmen in Committees*, Little Brown and Co., Boston, Massachusetts.

Ferejohn, J. (1974), *Pork Barrel Politics*, Stanford University Press, Stanford, California.

Fiorina, M. (1982), "Legislative Choice of Regulatory Forms: Legal Process or Administrative Process?," *Public Choice*, **39**, 33–66.

Florio, J. (1986), "Congress as Reluctant Regulator: Hazardous Waste Policy in the 1980's," *Yale Journal on Regulation*, **3**, 351–382.

Freeman, A. (1979), *The Benefits of Environmental Improvement*, Johns Hopkins University Press, Baltimore, Maryland.

Hahn, R. (1982), "Monitoring and the Choice of Instruments," Working Paper, School of Urban and Public Affairs, Carengie Mellon University, revised 1985.

Hahn, R. (1983), "Designing Markets in Transferable Property Rights: A Practitioner's Guide," in E. Joeres and M. David, eds., *Buying a Better Environment: Cost Effective Regulation Through Permit Trading*, University of Wisconsin Press, Madison, Wisconsin, 83–97.

Hahn, R. (1984), "Market Power and Transferable Property Rights," *Quarterly Journal of Economics*, **99**, 753–765.

Hahn, R. (1986), "Tradeoffs in Designing Markets with Multiple Objectives," *Journal of Environmental Economics and Management,* **13,** 1–12.

Hahn, R. (1987a) "A New Approach to the Design of Regulation in the Presence of Multiple Objectives," Working Paper 87–8, School of Urban and Public Affairs, Carnegie-Mellon University, forthcoming in the *Journal of Environmental Economics and Management.*

Hahn, R. (1987b), "Jobs and Environmental Qualtiy: Some Implications for Instrument Choice," *Policy Sciences,* **20,** 289–306.

Hahn, R. (1987c), "Rules, Equality and Efficiency: An Evaluation of Two Regulatory Reforms," Working Paper 87–7, School of Urban and Public Affairs, Carnegie-Mellon University, Pittsburgh, Pennsylvania.

Hahn, R. (1988), "An Evaluation of Options for Reducing Hazardous Waste," *Harvard Environmental Law Review,* **12,** 201–230.

Hahn, R. and Hester, G. (1986), "Where Did All the Markets Go?: An Analysis of EPA's Emission Trading Program," Working Paper 87–3. School of Urban and Public Affairs, Carnegie Mellon University. Pittsburgh. Pennsylvania, forthcoming in the *Yale Journal on Regulation.*

Hahn, R. and Hester, G. (1987a), "Marketable Permits: Lessons for Theory and Practice," Working Paper, School of Urban and Public Affairs, Carnegie Mellon University, Pittsburgh, Pennsylvania, forthcoming in the *Ecology Law Quarterly.*

Hahn, R. and Hester, G. (1987b), "The Maket for Bads: EPA's Experience with Emissions Trading," *Regulation,* **3/4,** 48–53.

Hahn, R. and McGartland, A. (1988), "The Political Economy of Instrument Choice: An Examination of the U.S. Role in Implementing the Montreal Protocol." Working Paper 88–34, School of Urban and Public Affairs. Carnegie Mellon University. Pittsburgh, Pennsylvania, forthcoming in the *Northwestern University Law Review.*

Hahn, R., McRae, G. and Milford, J. (1985), "Coping with Complexity in the Design of Environmental Policy," Working Paper 87–16, School of Urban and Public Affairs, Carnegie Mellon University. Pittsburgh, Pennsylvania, forthcoming in the *Journal of Environmental Management.*

Hahn, R. and Noll, R. (1982a), "Designing a Market for Tradable Emissions Permits." in W. Magat, ed., *Reform of Environmental Regulation.* Ballinger, Cambridge, Massachusetts, 119–146.

Hahn, R. and Noll, R. (1982b), "Implementing Tradable Emission Permits," in L. Graymer and F. Thompson, eds., *Reforming Social Regulation: Alternative Public Policy Strategies,* Sage Publications, Beverly Hills. California, 125–150.

Hahn, R. and Noll, R. (1983), "Barriers to Implementing Tradable Air Pollution Permits: Problems of Regulatory Interaction," *Yale Journal on Regulation,* **1,** 63–91.

Huber, P. (1987), "The Environmental Liability Dilemma," *CPCU Journal,* 206–216.

Hughes, J., Magat, W. and R. Williams, (1986). "The Economic Consequences of the OSHA Cotton Dust Standard: An Analysis of Stock Price Behavior," *Journal of Law and Economics, XXIX,* 29–59.

Kashmanian, R. *et al.* (1986). "Beyond Categorical Limits: The Case for Pollution Reduction Through Trading," paper presented at the 59th Annual Water Pollution Control Federation Conference, October 6–9, 35 pp.

Kneese, A. and Schultze, C. (1975). *Pollution, Prices, and Public Policy.* The Brookings Institution, Washington, D.C.

Lave, L. (1984), "Controlling Contradictions Among Regulations," *American Economic Review*, **74,** 471–475.

Levine, M. (1981), "Revisionism Revised? Airline Deregulation and the Public Interest," *Law and Contemporary Problems*, **44,** 179–195.

Lidgren, K. (1986), "Economic Instruments for Environmental Protection in Sweden," ENV/ECO/86.13, Organization for Economic Cooperation and Development, September 10, 40 pp.

Liroff, R. (1986), *Reforming Air Pollution Regulation: The Toil and Trouble of EPA's Bubble*, The Conservation Foundation, Washington, D.C.

Magat, W., Krupnick, A. and Harrington, W. (1986), *Rules in the Making: A Statistical Analysis of Regulatory Agency Behavior*, Resources for the Future, Washington, D.C.

Maloney, M. and McCormick, R. (1982), "A Positive Theory of Environmental Quality Regulation," *Journal of Law and Economics, XXV*, 99–123.

Mannix, B. (1987), Editor, *Regulation*, Interview, May 27.

Mayhew, D. (1974), *Congress: The Electrical Connection*, Yale Univesity Press, New Haven, Connecticut.

McCubbins, M. (1985), "The Legislative Design of Regulatory Structure," *American Journal of Political Science*, **29,** 721–748.

McCubbins, M. and Page, T. (1986), "The Congressional Foundations of Agency Performance," *Public Choice*, **51,** 173–190.

Meiburg, A. S. (1987), "Innovation and Infrastructure: Competing or Complementary Goals in Air Pollution Control," presented at the 80th Annual APCA Meetings, New York, New York, June 21–26.

Meidinger, E. (1985), "On Explaining the Development of 'Emissions Trading" in U.S. Air Pollution Regulation," *Law and Policy*, **7,** 447–479.

Melnick, R. (1983), *Regulation and the Courts*, Brookings Institution, Washington, D.C.

Mendeloff, J. (1988), *The Dilemma of Toxic Substance Regulation: How Overregulation Causes Underregulation at OSHA*, MIT Press, Cambridge, Massachusetts.

Menebroker, R. (1987), telephone interview, Chief, Project Review Branch, California Air Resources Board, Sacramento, California, July 21.

Montgomery, W. D. (1972), "Markets in Licenses and Efficient Pollution Control Programs," *Journal of Economic Theory*, **5,** 395–418.

National Academy of Sciences (1983), *Acid Deposition: Atmospheric Processes in Eastern North America*, National Research Council, National Academy Press, Washington, D.C.

National Clean Air Coalition (1985), *The Clean Air Act: A Briefing Book for the Members of Congress*, National Clean Air Coalition, Washington, D. C. April.

Niskanen, W. (1971). *Bureaucracy and Representative Government*. Aldine, Chicago, Illinois.

Noll, R. (1983), "The Political Foundations of Regulatory Policy." *Zeitschrift fur die gesamte Staatswissenschaft*, **139,** 377–404.

Novotny, G. (1986). "Transferable Discharge Permits for Water Pollution Control In Wisconsin," mimeo, December 1, 19 pp.

O'Neil, W. (1983), "The Regulation of Water Pollution Permit Trading under Conditions of Varying Streamflow and Temperature." in E. Joeres and M. David, eds., *Buying a Better Environment: Cost-Effective Regulation Through Permit Trading*, University of Wisconsin Press, Madison, Wisconsin, 219–231.

Opschoor, J. (1986). "Economic Instruments for Environmental Protection in the

Netherlands." ENV/ECO/86.15. Organization for Economic Cooperation and Development. August 1, 66 pp.

Overeynder, P. (1987), Telephone interview, Consultant to Northwest Colorado Council of Governments, Denver, Colorado, May 26.

Owen, B. and Braeutigam, R. (1978), *The Regulation Game*, Ballinger. Cambridge, Massachusetts.

Page, T. (1973), "Failure of Bribes and Standards for Pollution Abatement." *Natural Resources Journal,* **13,** 677–704.

Panella, G. (1986), "Economic Instruments for Environmental Protection in Italy," ENV/ECO/86.11, Organization for Economic Cooperation and Development, September 2, 42 pp.

Patterson, D. (1987), Telephone Interview, Bureau of Water Resources Management, Wisconsin Department of Natural Resources, Madison, Wisconsin, April 2.

Peltzman, S. (1976), "Toward a More General Theory of Regulation," *Journal of Law and Economics,* **19,** 211–240.

Pigou, A. (1932), *The Economics of Welfare,* Fourth Edition, Macmillan and Co., London.

Plott, C. (1983), "Externalities and Corrective Policies in Experimental Markets," *Economic Journal,* **93,** 106–127.

Portney, P. (1988), "Reforming Environmental Regulation: Three Modest Proposals." *Issues in Science and Technology. IV,* Winter, 74–81.

Roberts, M. and Spence, M. (1976), "Effluent Charges and Licenses Under Uncertainty," *Journal of Public Economics,* **5,** 193–208.

Rolph, E. (1983), "Government Allocation of Property Rights: Who Gets What?," *Journal of Policy Analysis and Management,* **3,** 45–61.

Russell, C., Harrington, W., and Vaughan, W. (1986), *Enforcing Pollution Control Laws,* Johns Hopkins University Press, Baltimore, Maryland.

Shepsle, K. and Weingast, B. (1984). "Political Solutions to Market Problems," *American Political Science Review,* **78,** 417–434.

Simon, H. (1976), *Administrative Behavior,* Third Edition, Free Press Press, Glencoe, Illinois.

Smith, F. (1986), "Superfund: A Hazardous Waste of Taxpayer Money," *Human Events,* August 2, 662–664, 671.

Sprenger, R. (1986), "Economic Instruments for Environemntal Protection in Germany," Organization for Economic Cooperation and Development. OECD, October 7, 78 pp.

Stewart, R. and J. Krier, J. (1978), *Environmental Law and Policy,* Bobbs-Merrill Co., New York.

Stigler, G. (1971), "The Theory of Economic Regulation," *Bell Journal of Economics,* **2,** 3–21.

Thomas, L. (1986), Memorandum Attached to Draft Emissions Trading Policy Statement, Environmental Protection Agency, Washington, D.C., May 19.

Thomas, L. (1987). "Testimony of Lee M. Thomas." Administrator, U.S. Environmental Protection Agency before the Subcommittee on Health and the Environment. Committee on Energy and Commerce, United States House of Representatives, February 19.

Tietenberg, T. (1985), *Emissions Trading: An Exercise in Reforming Pollution Policy,* Resources for the Future, Washington, D.C.

Tucker, W. (1982), *Progress and Privilege: America in the Age of Environmentalism.* Anchor Press, Garden City, New York.

U.S. Congressional Budget Office (1985), *Hazardous Waste Management: Recent Changes and Policy Alternatives,* May, U.S. Government Printing Office, Washington, D.C.

U.S. Environmental Protection Agency (1984), "Evaluation of Air Pollution Regulatory Strategies for Gasoline Marketing Industry." Office of Air and Radiation, EPA-450/3-84-012a, jUly.

U.S. Environmental Protection Agency (1985a). "Costs and Benefits of Reducing Lead in Gasoline, Final Regulatory Impact Analysis." Office of Policy Analysis. February.

U.S. Environmental Protection Agency (1985b). " Quarterly Reports on Lead in Gasoline." Field Operations and Support Division. Office of Air and Radiation. July 16.

U.S. Environmental Protection Agency (1986), "Quarterly Reports on Lead in Gasoline," Field Operations and Support Division, Office of Air and Radiation, March 21, May 23, July 15.

Wall Street Journal (1986), "Superfund Bill Cleared By House Faces Reagan Veto," by Robert E. Taylor, Midwest edition, October 9, p. 22.

Weingast, B. (1981), "Regulation, Reregulation, and Deregulation: The Political Foundations of Agency Clientele Relationships," *Law and Contemporary Problems,* **44,** 147–177.

Weitzman, M. (1974), "Prices vs. Quantities," *Review of Economic Studies.* **41,** 477–491.

Welch, W. (1983), "The Political Feasibility of Full Ownership Property Rights: The Cases of Pollution and Fisheries," *Policy Sciences,* **16,** 165–180.

Wilson, J., ed. (1980), *The Politics of Regulation,* Basic Books, Inc., New York.

Yohe, G. (1976), "Polluters' Profits and Political Response: Direct Control Versus Taxes: Comment," *American Economic Review,* **66,** 981–982.

Index

131

132INDEX

FUNDAMENTALS OF PURE AND APPLIED ECONOMICS

SECTIONS AND EDITORS

BALANCE OF PAYMENTS AND INTERNATIONAL FINANCE
W. Branson, Princeton University

DISTRIBUTION
A. Atkinson, London School of Economics

ECONOMIC DEMOGRAPHY
T.P. Schultz, Yale University

ECONOMIC DEVELOPMENT STUDIES
S. Chakravarty, Delhi School of Economics

ECONOMIC FLUCTUATIONS: FORECASTING, STABILIZATION, INFLATION, SHORT TERM MODELS, UNEMPLOYMENT
A. Ando, University of Pennsylvania

ECONOMIC HISTORY
P. David, Stanford University, and M. Lévy-Leboyer, Université Paris X

ECONOMIC SYSTEMS
J.M. Montias, Yale University, and J. Kornai, Institute of Economics, Hungarian Academy of Sciences

ECONOMICS OF HEALTH, EDUCATION, POVERTY AND CRIME
V. Fuchs, Stanford University

ECONOMICS OF THE HOUSEHOLD AND INDIVIDUAL BEHAVIOR
J. Muellbauer, University of Oxford

ECONOMICS OF TECHNOLOGICAL CHANGE
F. M. Scherer, Swarthmore College

ECONOMICS OF UNCERTAINTY AND INFORMATION
S. Grossman, Princeton University, and J. Stiglitz, Princeton University

EVOLUTION OF ECONOMIC STRUCTURES, LONG-TERM MODELS, PLANNING POLICY, INTERNATIONAL ECONOMIC STRUCTURES
W. Michalski, O.E.C.D., Paris

EXPERIMENTAL ECONOMICS
C. Plott, California Institute of Technology

GAME THEORY
R. Aumann, The Hebrew University of Jerusalem

GENERAL EQUILIBRIUM THEORY AND OPTIMUM THEORY
W. Hildenbrand, University of Bonn, and A. Mas-Colell, Harvard University

GOVERNMENT OWNERSHIP AND REGULATION OF ECONOMIC ACTIVITY
E. Bailey, Carnegie-Mellon University

INTERNATIONAL ECONOMIC ISSUES
T. Fujii, University of Nagoya

INTERNATIONAL TRADE
M. Kemp, University of New South Wales

LABOR ECONOMICS
F. Welch, University of California, Los Angeles

LAW AND ECONOMICS
S. Shavell, Harvard Law School

MACROECONOMIC THEORY
J. Grandmont, CEPREMAP

MARXIAN ECONOMICS
J. Roemer, University of California, Davis

MONETARY THEORY
N. Wallace, University of Minnesota

NATURAL RESOURCES AND ENVIRONMENTAL ECONOMICS
C. Henry, Ecole Polytechnique, Paris

ORGANIZATION THEORY AND ALLOCATION PROCESSES
A. Postlewaite, University of Pennsylvania, and D. Schmeidler, Tel Aviv University

POLITICAL SCIENCE AND ECONOMICS
J. Ferejohn, Stanford University

PROGRAMMING METHODS IN ECONOMICS
M. Balinski, Ecole Polytechnique, Paris

PUBLIC EXPENDITURES
P. Dasgupta, University of Oxford

REGIONAL AND URBAN ECONOMICS
R. Arnott, Queen's University at Kingston

SOCIAL CHOICE THEORY
A. Sen, University of Oxford

TAXES
R. Guesnerie, Ecole des Hautes Etudes en Sciences Sociales

THEORY OF ECONOMIC GROWTH
J. Scheinkman, University of Chicago

THEORY OF THE FIRM AND INDUSTRIAL ORGANIZATION
A. Jacquemin, Université Catholique de Louvain

PUBLISHED TITLES

ISSN: 0191-1708